DEFENSE, SECURITY AND STRATEGIES

FEDERAL INFORMATION SECURITY

SELECT PROTECTION AND CONTROL EFFORTS

DEFENSE, SECURITY AND STRATEGIES

Additional books in this series can be found on Nova's website under the Series tab.

Additional E-books in this series can be found on Nova's website under the E-books tab.

SAFETY AND RISK IN SOCIETY

Additional books in this series can be found on Nova's website under the Series tab.

Additional E-books in this series can be found on Nova's website under the E-books tab.

DEFENSE, SECURITY AND STRATEGIES

FEDERAL INFORMATION SECURITY

SELECT PROTECTION AND CONTROL EFFORTS

DAVID G. MILLER
AND
IVAN I. WILSON
EDITORS

Nova Science Publishers, Inc.
New York

For permission to use material from this book please contact us:
Telephone 631-231-7269; Fax 631-231-8175
Web Site: http://www.novapublishers.com

Library of Congress Cataloging-in-Publication Data

ISBN 978-1-61942-771-6

Published by Nova Science Publishers, Inc. † New York

CONTENTS

PREFACE

In the last several years, federal agencies have increasingly adopted the use of wireless technologies. While wireless technologies provide many potential benefits, including greater flexibility for a mobile workforce and ease of installation and use, they also pose significant risks to information and systems. Wireless technologies use radio waves instead of direct physical connections to transmit data between networks and devices. As a result, without proper security precautions, these data can be more easily intercepted and altered than if being transmitted through physical connections. This book examines the select protection and control efforts being utilized to secure federal agency wireless networks, the FDIC; and the national nuclear administration.

Chapter 1 - Over the past several years, federal agencies have rapidly adopted the use of wireless technologies for their information systems. In a 2005 report, GAO recommended that the Office of Management and Budget (OMB), in its role overseeing government wide information security, take several steps to help agencies better secure their wireless networks.

GAO was asked to update its prior report by (1) identifying leading practices and state-of-the-art technologies for deploying and monitoring secure wireless networks and (2) assessing agency efforts to secure wireless networks, including their vulnerability to attack.

To do so, GAO reviewed publications, guidance, and other documentation and interviewed subject matter experts in wireless security. GAO also analyzed policies and plans and interviewed agency officials on wireless security at 24 major federal agencies and conducted additional detailed testing at these 5 agencies: the Departments of Agriculture, Commerce, Transportation, and Veterans Affairs, and the Social Security Administration.

Chapter 2 - The Federal Deposit Insurance Corporation (FDIC) has a demanding responsibility enforcing banking laws, regulating financial institutions, and protecting depositors. Because of the importance of its work, the corporation must employ strong information security controls to ensure that its information systems are adequately protected from inadvertent misuse, fraud, and improper disclosure.

As part of its audit of the 2009 financial statements of the Deposit Insurance Fund and the Federal Savings & Loan Insurance Corporation Resolution Fund administrated by FDIC, GAO assessed (1) the effectiveness of FDIC's controls in protecting the confidentiality, integrity, and availability of its financial systems and information and (2) the progress FDIC has made in mitigating previously reported information security weaknesses. To perform the audit, GAO examined security policies, procedures, reports, and other documents; tested controls over key financial applications; and interviewed key FDIC personnel.

Chapter 3 - In the absence of underground nuclear weapons testing, the National Nuclear Security Administration (NNSA) relies on its supercomputing operations at its three weapons laboratories to simulate the effects of changes to current weapons systems, calculate the confidence of future untested systems, and ensure military requirements are met.

GAO was requested to assess the extent to which (1) NNSA has implemented contingency and disaster recovery planning and testing for its classified supercomputing systems, (2) the laboratories are able to share supercomputing capacity for recovery operations, and (3) NNSA tracks the costs for contingency and disaster recovery planning for supercomputing assets. To do this work, GAO examined contingency and disaster recovery planning policies and activities, and analyzed classified supercomputing capabilities at the weapons laboratories, and NNSA budgetary data.

Chapter 4 - To reduce the threat to federal systems and operations posed by cyber attacks on the United States, the Office of Management and Budget (OMB) launched, in November 2007, the Trusted Internet Connections (TIC) initiative, and later, in 2008, the Department of Homeland Security's (DHS) National Cybersecurity Protection System (NCPS), operationally known as Einstein, became mandatory for federal agencies as part of TIC. For each of these initiatives, GAO was asked to (1) identify their goals, objectives, and requirements; (2) determine the status of actions federal agencies have taken, or plan to take, to implement the initiatives; and (3) identify any benefits, challenges, and lessons learned. To do this, GAO reviewed plans, reports, and other documents at 23 major executive branch agencies, interviewed officials, and reviewed OMB and DHS guidance.

In: Federal Information Security
Editors: D. G. Miller and I. I. Wilson

ISBN 978-1-61942-771-6

Chapter 1

INFORMATION SECURITY: FEDERAL AGENCIES HAVE TAKEN STEPS TO SECURE WIRELESS NETWORKS, BUT FURTHER ACTIONS CAN MITIGATE RISK

United States Government Accountability Office

WHY GAO DID THIS STUDY

Over the past several years, federal agencies have rapidly adopted the use of wireless technologies for their information systems. In a 2005 report, GAO recommended that the Office of Management and Budget (OMB), in its role overseeing governmentwide information security, take several steps to help agencies better secure their wireless networks.

GAO was asked to update its prior report by (1) identifying leading practices and state-of-the-art technologies for deploying and monitoring secure wireless networks and (2) assessing agency efforts to secure wireless networks, including their vulnerability to attack.

To do so, GAO reviewed publications, guidance, and other documentation and interviewed subject matter experts in wireless security. GAO also analyzed policies and plans and interviewed agency officials on wireless security at 24 major federal agencies and conducted additional detailed testing at these 5 agencies: the Departments of Agriculture, Commerce, Transportation, and Veterans Affairs, and the Social Security Administration.

WHAT GAO RECOMMENDS

GAO is making two recommendations to OMB to enhance governmentwide oversight and four recommendations to the Department of Commerce for additional guidelines related to wireless security. The Department of Commerce concurred with GAO's recommendations. OMB did not provide comments on the report.

WHAT GAO FOUND

GAO identified a range of leading security practices for deploying and monitoring secure wireless networks and technologies that can help secure these networks. The leading practices include the following:

- comprehensive policies requiring secure encryption and establishing usage restrictions, implementation practices, and access controls;
- a risk-based approach for wireless deployment and monitoring;
- a centralized wireless management structure that is integrated with the management of the existing wired network;
- configuration requirements for wireless networks and devices;
- incorporation of wireless and mobile device security in training;
- use of encryption, such as a virtual private network for remote access;
- continuous monitoring for rogue access points and clients; and
- regular assessments to ensure wireless networks are secure.

Agencies have taken steps to secure their wireless networks, but more can be done to improve security and to limit vulnerability to attack. Specifically, application was inconsistent among the agencies for most of the following leading practices:

- Most agencies developed policies to support federal guidelines and leading practices, but gaps existed, particularly with respect to dual-connected laptops and mobile devices taken on international travel.
- All agencies required a risk-based approach for management of wireless technologies.

- Many agencies used a decentralized structure for management of wireless, limiting the standardization that centralized management can provide.
- The five agencies where GAO performed detailed testing generally securely configured wireless access points but had numerous weaknesses in laptop and smartphone configurations.
- Most agencies were missing key elements related to wireless security in their security awareness training.
- Twenty agencies required encryption, and eight of these agencies specified that a virtual private network must be used; four agencies did not require encryption for remote access.
- Many agencies had insufficient practices for monitoring or conducting security assessments of their wireless networks.

Existing governmentwide guidelines and oversight efforts do not fully address agency implementation of leading wireless security practices. Until agencies take steps to better implement these leading practices, and OMB takes steps to improve governmentwide oversight, wireless networks will remain at an increased vulnerability to attack.

ABBREVIATIONS

DISA	Defense Information Systems Agency
DHS	Department of Homeland Security
EAP	extensible authentication protocol
FDCC	Federal Desktop Core Configuration
FIPS	Federal Information Processing Standards
FISMA	Federal Information Security Management Act
IEEE	Institute of Electrical and Electronics Engineers
IT	information technology
IPv6	Internet protocol version 6
NIST	National Institute of Standards and Technology
OMB	Office of Management and Budget
PDA	personal digital assistant
SP	Special Publications
VPN	virtual private network
WEP	wired equivalent privacy
WiMAX	Worldwide Interoperability for Microwave Access

WPA	Wi-Fi Protected Access
WLAN	wireless local area network

November 30, 2010

The Honorable Richard J. Durbin
Chairman
The Honorable Susan Collins
Ranking Member
Subcommittee on Financial Services and General Government
Committee on Appropriations
United States Senate

The Honorable José E. Serrano
Chairman
The Honorable Jo Ann Emerson
Ranking Member
Subcommittee on Financial Services and General Government
Committee on Appropriations
House of Representatives

In the last several years, federal agencies have increasingly adopted the use of wireless technologies. While wireless technologies provide many potential benefits, including greater flexibility for a mobile workforce and ease of installation and use, they also pose significant risks to information and systems. Wireless technologies use radio waves instead of direct physical connections to transmit data between networks and devices. As a result, without proper security precautions, these data can be more easily intercepted and altered than if being transmitted through physical connections.

We have previously reported on the security of wireless networks at federal agencies in 2005.[1] The conference report accompanying the Financial Services and General Government Appropriations Act, 2010, directed us to update our 2005 report.[2] Accordingly, our objectives for this report were to: (1) identify leading practices and state-of-the-art technologies for deploying and monitoring secure wireless networks and (2) assess agency efforts to secure wireless networks, including their vulnerability to attack.

To identify leading practices and state-of-the-art technologies for deploying and monitoring secure wireless networks, we obtained and reviewed publications, guidance, and other documentation, and interviewed private and federal subject matter experts. To assess agency efforts to secure wireless networks, we reviewed agency documents and conducted structured interviews with agency officials to learn about the wireless posture at 24 major federal agencies.[3] We supplemented these questions with site visits and detailed testing of wireless security controls at five of the agencies.

We conducted this performance audit from January 2010 to November 2010, in accordance with generally accepted government auditing standards. Those standards require that we plan and perform the audit to obtain sufficient, appropriate evidence to provide a reasonable basis for our findings and conclusions based on our audit objectives. We believe that the evidence obtained provides a reasonable basis for our findings and conclusions based on our audit objectives. Appendix I contains additional details on the objectives, scope, and methodology of our review.

BACKGROUND

The advantages of wireless technology for federal agencies include increased flexibility, easier installation, and easier scalability than wired technologies. If a federal agency has installed a wireless infrastructure, users with wireless-enabled devices can more easily connect to the agency's network throughout its facilities. In addition, agency employees traveling with wireless-enabled devices may be able to connect to an agency network via any one of the many public Internet access points or hot spots. Installation can be easier and less costly because the network can be established without having to pull cables through walls or ceilings or modify the physical network infrastructure. Wireless networks can also be easily scaled from small peer-to-peer networks to very large enterprise networks. For example, an agency can greatly expand the size of its wireless network and the number of users it can serve by increasing the number of access points. The following wireless technologies are commonly used by federal agencies:

- *wireless local area network (WLAN)*—a group of wireless networking nodes within a limited geographic area that serve as an extension to existing wired local area networks;
- *wireless personal area network*—used to establish small-scale wireless networks such as those using Bluetooth®, which is an open standard for short-range communication; and

- *wireless cellular networks*—a telecommunications network managed by a service provider that supports smartphones, which offer the ability to provide data such as e-mail and Web browsing wirelessly over cellular networks, and cellular data cards, which provide Internet connectivity to laptop computers.

Wireless Local Area Networks

WLANs are generally composed of two basic elements: access points and other wireless-enabled client devices, such as laptop computers. These elements rely on radio transmitters and receivers to communicate with each other. Access points are physically wired to a conventional network and provide a means for wireless devices to connect to them. WLANs that are based on the Institute of Electrical and Electronics Engineers[4] (IEEE) 802.11 standards are also known as Wi-Fi.

WLANs are characterized by one of the following two basic structures, referred to as infrastructure mode and ad hoc mode:

- *Infrastructure mode.* By deploying one or more access points that broadcast overlapping signals, an organization can achieve broad wireless network coverage. Infrastructure mode enables a laptop or other mobile device to be moved about freely while maintaining access to the resources of the wired network (see figure 1).

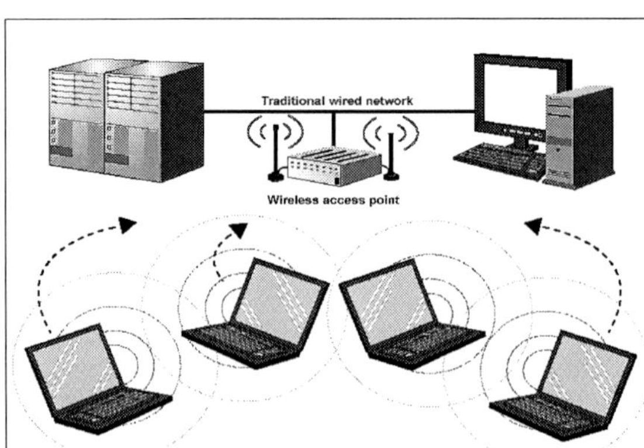

Sources: GAO; Microsoft Visio and Art Explosion (images).

Figure 1. Example of a Wireless Infrastructure Mode Network.

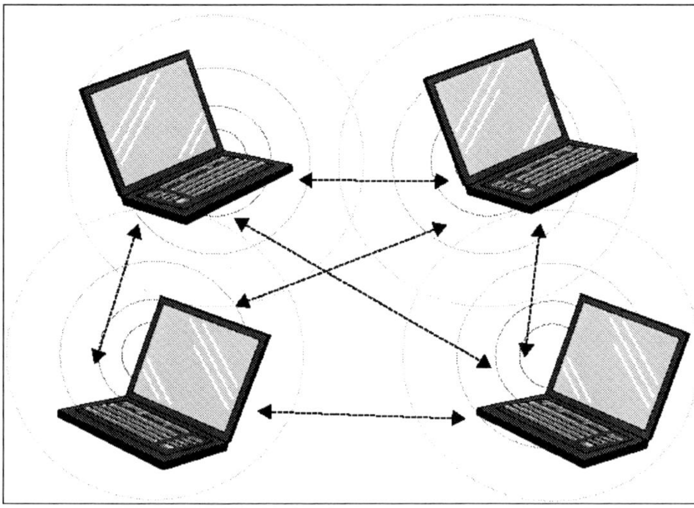

Sources: GAO; Microsoft Visio and Art Explosion (images).

Figure 2. Example of Wireless Ad Hoc Networking.

- *Ad hoc mode.* This type of wireless structure allows wireless devices that are near one another to easily interconnect. In ad hoc mode, wireless- enabled devices can share network functionality without the use of an access point or a wired network connection (see figure 2).

After approval of the initial IEEE 802.11 standard in 1997, IEEE released several 802.11 amendments to increase WLAN network speeds to be more comparable to that of wired networks. The 802.11 standard and these subsequent amendments include security features known collectively as wired equivalent privacy (WEP). However, configurations that use WEP have significant security flaws.

To address these flaws, IEEE released the 802.11i security standard in 2004, which specifies security components that work together with 802.11 transmission standards. The IEEE 802.11i security standard supports wireless connections that provide moderate to high levels of assurance against WLAN security threats through the use of different cryptographic techniques.

While IEEE was developing 802.1 1i, the Wi-Fi Alliance[5] developed the Wi-Fi Protected Access (WPA) security certification as an interim means to improve security over WEP. The protocols used under the WPA certification address vulnerabilities of WEP, but the certification does not require support for strong encryption.

In conjunction with the ratification of the 802.11i security standard in 2004, the Wi-Fi Alliance introduced WPA2—the interoperability certification for 802.11i. The WPA2 certification extends the security capabilities offered by WPA to include all requirements of the 802.11i standard. Both WPA and WPA2 offer two modes of operation: Personal and Enterprise. WPA2-Personal protects unauthorized network access by using a preshared password as a key for network setup and access, while WPA2-Enterprise verifies network users through an authentication server. In most cases, WPA2-Enterprise is recommended to eliminate the continuous process of generating, deploying, and replacing outdated passwords. Although WPA2-Enterprise-certified products provide more security protections than WEP and WPA, recent reports revealed that wireless networks protected with WPA2-Enterprise encryption can also be susceptible to attacks.

Most recently, in 2009, IEEE ratified the 802.1 1w-2009 standard, which further increases the overall security of 802.11–based networks. Specifically, 802.11w-2009 provides improved protection for WLANs by defining additional encryption security features to help prevent incidents such as denial of service attacks against WLANs.

Wireless Personal Area Networks

Wireless personal area networks provide wireless connectivity to devices such as telephone headsets or computer keyboards within close proximity. Bluetooth is commonly used to establish these types of networks. Several versions of the Bluetooth standard have been adopted by the Bluetooth Special Interest Group.[6]

Each Bluetooth device must operate in one of the four security modes defined by the Bluetooth standard. Each version of Bluetooth supports some, but not all, of these modes.

Wireless Cellular Networks

Cellular networks are managed by service providers who provide coverage based on dividing a large geographical service area into smaller areas of coverage called cells. As a mobile phone moves from one cell to another, a cellular arrangement requires active connections to be monitored and effectively passed along between cells to maintain the connection.

In addition to cellular phones, cellular networks support smartphones and cellular data cards. Smartphones offer more functionality than basic cellular phones, including e-mail and other office productivity applications and have extended expansion capabilities through peripheral card slots and other built-in wireless communications such as Bluetooth and Wi-Fi. Cellular data cards allow laptop users to connect to the Internet anywhere cellular service is available. However, cellular data cards can only access the Internet if the user is within the service provider's network coverage area.

Federal Agencies Make Widespread Use of Wireless Technologies

Agencies reported significant use of WLANs to extend working mobility for employees and contractors. For example, 18 agencies reported using WLANs in a variety of ways. Five agencies reported having wireless networks available for headquarters along with field offices or components. Twelve other agencies reported that components have different wireless practices than headquarters. For example, one major agency reported no WLANs at its headquarters, but it has components that use them. Further, several agencies use wireless networks for more limited purposes than connecting to the core agency network. Specifically, five agencies reported offering WLANs that connect directly to the Internet for use in conference rooms or other public spaces. Another agency reported using wireless access points to provide Internet connectivity at outdoor construction sites.

Personal area networks using Bluetooth technology were also reported by many agencies. Specifically, 14 agencies reported using Bluetooth devices. Ten agencies reported permitting cellular phone users to connect wireless headsets, and four agencies reported permitting wireless keyboards or mice.

Agencies also reported extensive use of smartphones and cellular data cards. All 24 agencies we queried reported using smartphones, primarily the BlackBerry® brand. Agencies' smartphone management structures included: management through a central server located at the department level or at the component level and one component or office providing smartphone management to another office. Seventeen agencies reported using cellular data cards to provide Internet connectivity to user laptops.

These cards and services are typically provided by commercial telecommunications carriers.

Table 1. Examples of Network Security Threats

Denial-of-service	Preventing or limiting the normal use or management of networks or network devices.
Eavesdropping	Passively monitoring network communications for data, including authentication credentials.
Man-in-the-middle	Actively impersonating multiple legitimate parties, such as appearing as a client to an access point and appearing as an access point to a client. Allows attacker to intercept communications between an access point and a client, thereby obtaining authentication credentials and data.
Masquerading	Impersonating an authorized user and gaining unauthorized privileges.
Message modification	Altering a legitimate message by deleting, adding to, changing, or reordering it.
Message replay	Passively monitoring transmissions and retransmitting messages, acting as if the attacker were a legitimate user.
Misappropriation	Stealing or making unauthorized use of a service.
Traffic analysis	Passively monitoring transmissions to identify communication patterns and participants.

Source: GAO analysis of NIST data.

Wireless Technologies Are Susceptible to Security Risks

Without proper safeguards, computer systems are vulnerable to individuals and groups with malicious intent who can intrude and use their access to obtain sensitive information, commit fraud, disrupt operations, or launch attacks against other computer systems and networks. The risk to these systems is well-founded for a number of reasons, including the dramatic increase in reports of security incidents, the ease of obtaining and using hacking tools, and steady advances in the sophistication and effectiveness of attack technology. Table 1 provides a compilation of threats to wireless and wired networks as identified by the National Institute of Standards and Technology (NIST).

Wireless networks also face challenges that are unique to their environment. A significant difference between wireless and wired networks is the relative ease of intercepting WLAN transmissions. For WLANs, attackers only need to be in range of wireless transmissions and do not have to gain physical access to the network or remotely compromise systems on the network. WLANs also have to protect against the deployment of unauthorized wireless devices, such as access points, that are configured to appear as part of an agency's wireless network infrastructure. In implementing wireless networks, federal agencies need to address these challenges to maintain the confidentiality, integrity, and availability of the information.

Bluetooth-enabled devices are susceptible to general networking threats and are also threatened by more specific Bluetooth-related attacks such as bluesnarfing, which enables attackers to gain access to a Bluetoothenabled device by exploiting a software flaw in older devices.

Smartphones are also susceptible to general networking threats and face additional security risks. Those risks include those caused by their size and portability, as well as the availability of different wireless interfaces and associated services. For example, the size and portability of smartphones can result in the loss of physical control of a device that could reveal sensitive data to an unauthorized user.

Recent articles released by the media reinforce the need for federal agencies to secure their wireless networks and devices. Examples of reported incidents and risks include the following:

- A retail company admitted in 2007 that hackers located and tested wireless networks for vulnerabilities and installed programs on these networks to steal the credit card information of more than 45 million consumers.
- An assessment of wireless vulnerability conducted in 2008 at 27 airports that had wireless networks found that personal information could be leaked because only 3 percent of hot spot users used a virtual private network (VPN)[7] to encrypt their data.
- In 2009, a counterintelligence official described how smartphones could have been tagged, tracked, monitored, and exploited at the 2008 Beijing Olympics. The malicious software could have also posed a threat to e-mail servers in the United States.

Scenarios Provide Examples of Attacks Using Wireless Vulnerabilities

The following scenarios (figures 3-5) provide examples of well-known attacks used to exploit vulnerabilities in wireless technologies. These scenarios do not represent all possible attacks on wireless technology vulnerabilities.

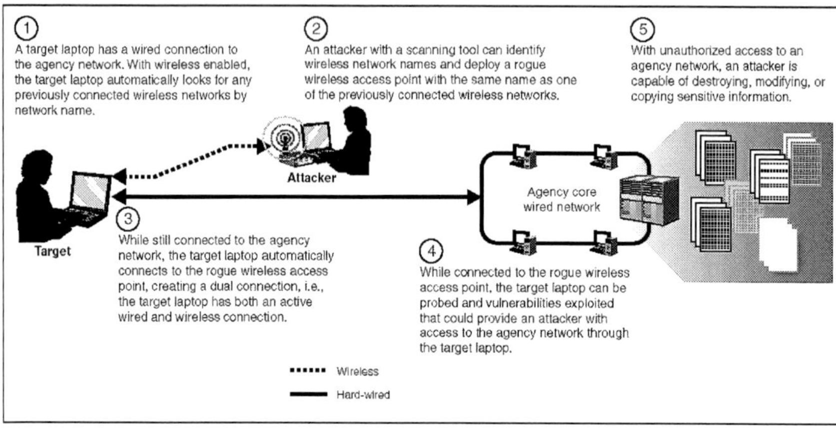

Source: GAO; Art Explosion (images).

Figure 3. Dual-Connect Attack Scenario.

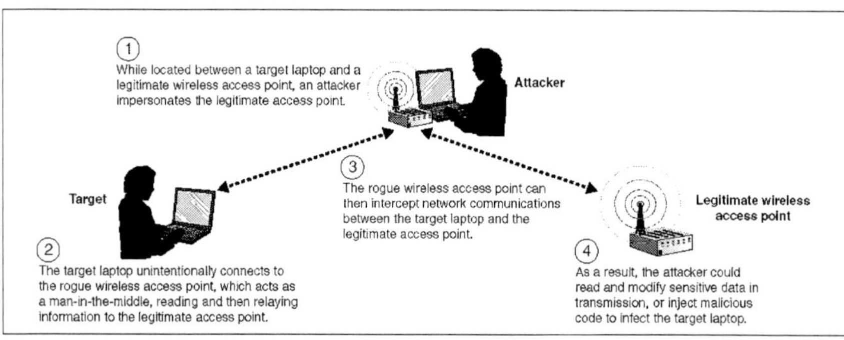

Source: GAO; Art Explosion (images).

Figure 4. Wireless Man-in-the-Middle Attack Scenario.

In a dual-connect scenario (see figure. 3), the attacker exploits insecure laptop configurations to gain unauthorized access to an organization's core network.

Wireless man-in-the-middle attacks (see figure 4) use an insecure laptop configuration to intercept or alter information transmitted wirelessly between the target laptop and a wireless access point.

Attacks on smartphones (see figure 5) can involve stealing data or injecting malicious code using phone storage cards.

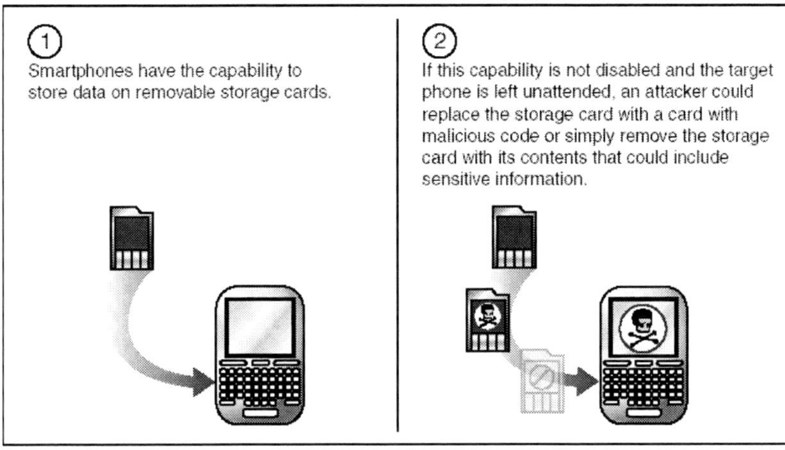

Source: GAO.

Figure 5. Smartphone Data Attack Scenario.

Federal Laws and Guidelines Provide a Framework for Wireless Security Policies

The Federal Information Security Management Act (FISMA) of 2002 requires each agency to develop, document, and implement an agencywide information security program to provide security for the data and information systems that support the agency's operations and assets.[8] Significant amounts of agency data are stored on and transmitted through wireless devices and networks. Wireless technologies are often important parts of the information systems that support the agency's operations and assets. Accordingly, wireless technologies are typically encompassed by agency information security programs required under FISMA. FISMA also assigns additional information security responsibilities for the Office of Management and Budget (OMB) and NIST.

FISMA assigns OMB specific responsibilities, including

- overseeing the implementation of policies, standards, and guidelines on information security, including ensuring timely agency adoption of and compliance with standards;
- requiring agencies to identify and provide information security protections commensurate with the risk and magnitude of the harm resulting from the unauthorized access, use, disclosure, disruption,

modification, or destruction of information collected or maintained by or on behalf of an agency, or information systems used or operated by or on behalf of an agency;

- overseeing agency compliance with FISMA requirements;
- reviewing at least annually, and approving or disapproving, agency information security programs; and
- annual reporting to Congress on agency compliance with the requirements of FISMA, including significant deficiencies in agency information security practices and planned remedial action to address such deficiencies.

In a July 2010 memo, OMB directed the Department of Homeland Security (DHS) to exercise primary responsibility within the executive branch for the operational aspects of federal agency cybersecurity with respect to the federal information systems that fall within FISMA.[9] According to the memo, DHS is to oversee the implementation of and reporting on information security policies and guidance in federal agencies, oversee agency compliance with FISMA, and annually review agency cybersecurity programs. OMB will continue to report annually to Congress on the progress of agencies' compliance with FISMA.

According to the Director of Federal Network Security—the DHS official responsible for many of DHS's newly assigned FISMA-related activities—DHS is beginning its oversight activities through the annual FISMA reporting process that federal agencies are required to follow. The official stated that the department does not currently have any wireless-securityspecific activities under way, but that the department is planning future activities that may address wireless security, including compliance audits and an architecture document.

Under FISMA, NIST is responsible for developing standards and guidelines that include minimum information security requirements. Table 2 describes NIST Special Publications (SP) that include guidelines intended to secure wireless technologies.

NIST is also responsible for administering the United States Configuration Baseline, which is an initiative to create security configuration baselines for information technology (IT) products deployed across federal agencies.

In addition to NIST guidelines, other federal agencies have developed guidance for securing wireless technologies. For example, the Department of Defense's Defense Information Systems Agency (DISA) has created a series of security technical implementation guides that address general purpose or multiuse technologies. These guides serve as configuration standards for the Department of Defense's wireless devices and systems. In addition, DISA has made these guides available for other federal agencies to provide them with a baseline level of security.

Table 2. Wireless Security Guidelines Identified in NIST Guidelines

NIST SP	Purpose
800-48, *Guide to Securing Legacy IEEE 802.11 Wireless Networks*[a]	Provides guidelines to organizations in securing their legacy IEEE 802.11 WLAN that cannot use the IEEE 802.11i standard.
800-53, *Recommended Security Controls for Federal Information Systems and Organizations*[b]	Provides guidelines for selecting and specifying security controls for information systems that include wireless access and access controls for mobile devices.
800-94, *Guide to Intrusion Detection and Prevention Systems*[c]	Provides a basis for designing, implementing, configuring, securing, monitoring, and maintaining intrusion detection and prevention systems including a wireless intrusion detection system.[d]
800-97, *Establishing Wireless Robust Security Networks: A Guide to IEEE 802.11i*[e]	Assists organizations in understanding, selecting, and implementing technologies based on IEEE 802.11i.
800-101, *Guidelines on Cell Phone Forensics*[f]	Provides basic information on the preservation, acquisition, examination, analysis, and reporting of digital evidence on cell phones, relevant to law enforcement, incident response, and other types of investigations.
800-114, *User's Guide to Securing External Devices for Telework and Remote Access*[g]	Provides guidelines for securing external devices used for telework including wireless home networks and wireless-enabled personal computers.
800-120, *Recommendation for EAP Methods Used in Wireless Network Access Authentication*[h]	Formalizes a set of core security requirements for extensible authentication protocol (EAP)[i] for wireless access authentication and key establishment.
800-121, *Guide to Bluetooth Security*[j]	Provides information on the security capabilities of Bluetooth and provides recommendations to secure Bluetooth devices effectively.

Table 2. (Continued)

NIST SP	Purpose
800-124, *Guidelines on Cell Phone and PDA Security*[k]	Provides an overview of cell phone and personal digital assistant (PDA) devices in use today and provides safeguards for securing these devices.

Source: GAO analysis of NIST data.

[a] NIST, *Guide to Securing Legacy IEEE 802.11 Wireless Networks,* SP 800-48 Revision 1 (Gaithersburg, MD: July 2008).

[b] NIST, *Recommended Security Controls for Federal Information Systems and Organizations,* SP 800- 53 Revision 3 (Gaithersburg, MD: August 2009).

[c] NIST, *Guide to Intrusion Detection and Prevention Systems (IDPS),* SP 800-94 (Gaithersburg, MD: February 2007).

[d] An intrusion detection system monitors the events occurring in a computer system or network and analyzes them for signs of possible incidents.

[e] NIST, *Establishing Wireless Robust Security Networks: A Guide to IEEE 802.11i,* SP 800-97 (Gaithersburg, MD: February 2007).

[f] NIST, *Guidelines on Cell Phone Forensics,* SP 800-1 01 (Gaithersburg, MD: May 2007).

[g] NIST, *User's Guide to Securing External Devices for Telework and Remote Access,* SP 800-114 (Gaithersburg, MD: November 2007).

[h] NIST, *Recommendation for EAP Methods Used in Wireless Network Access Authentication,* SP 800- 120 (Gaithersburg, MD: September 2009).

[i] EAP supports multiple authentication methods used when connecting a computer to the Internet.

[j] NIST, *Guide to Bluetooth Security,* SP 800-121 (Gaithersburg, MD: September 2008).

[k] NIST, *Guidelines on Cell Phone and PDA Security,* SP 800-124 (Gaithersburg, MD: October 2008).

GAO Has Previously Recommended Improvements to Wireless Network Security Guidance

In 2005, we reported that federal agencies lacked key controls for securing wireless networks.[10] We recommended that the Director of OMB instruct federal agencies to ensure that wireless network security is incorporated into their agencywide information security programs, in accordance with FISMA.

Specifically, we recommended that agencywide security programs should include the following security controls.

- Robust policies for authorizing the use of the wireless networks, identifying requirements, and establishing security controls for wireless- enabled devices in accordance with NIST guidelines.
- Security configuration requirements for wireless devices that include
  - security tools, such as encryption, authentication, VPN, and firewalls;
  - placement and strength of wireless access points to minimize signal leakage; and
  - physical protection of wireless-enabled devices.
- Comprehensive monitoring programs, including the use of tools such as site surveys and intrusion detection systems to
  - detect signal leakage;
  - ensure compliance with configuration requirements;
  - ensure only authorized access and use of wireless networks; and
  - identify unauthorized wireless-enabled devices and activities in the agency's facilities.
- Wireless security training for employees and contractors.

Table 3. Guidelines in NIST Publications Addressing Recommendations from GAO-05-383

Recommendation	NIST SP 800-48	NIST SP 800-53	NIST SP 800-97	Area
Establish policies	X	X	X	Establishing wireless networking security policies, such as infrastructure and client device security; criteria for identifying and implementing security requirements; access controls for portable and mobile devices; and establishing and maintaining robust security for wireless local area networks.

Table 3. (Continued)

Recommendation	NIST SP 800-48	NIST SP 800-53	NIST SP 800-97	Area
Configuration requirements include security tools	X	X		Configuring wireless client device security tools and the use of security tools such as personal firewalls, host-based intrusion detection and prevention systems for the protection of wireless
Configuration requirements address access points	X			Establishing access point configuration and awareness of access point security concerns, including signal boundary considerations.
Configuration requirements include physical protection	X	X		Ensuring physical protection of wireless devices such as usage restrictions and implementation guidance for organization-controlled portable and mobile devices.
Monitoring programs include tools to detect signal leakage	X			Determining criteria for conducting site surveys and the use of appropriate wall-mounted antennas to minimize signal leakage.
Monitoring programs include tools to ensure configuration compliance	X	X		Using wireless intrusion detection and prevention systems to determine misconfigured clients and using policy driven software solutions to ensure client devices and users comply with defined WLAN policies.
Monitoring programs include tools to ensure access is authorized	X	X	X	Using wireless intrusion detection and prevention systems to determine whether unauthorized users or devices are attempting to access, have already accessed, or have compromised the WLAN; and performing regular audits using wireless sniffers and other tools to determine whether wireless products are transmitting correctly and on the correct channels.

Recommendation	NIST SP 800-48	NIST SP 800-53	NIST SP 800-97	Area
Monitoring programs include tools to identify unauthorized access	X	X	X	Using wireless intrusion detection and prevention systems to detect suspicious or unauthorized activity and completing site surveys to discover any sources of radio interference.
Security training	X	X	X	Establishing wireless security awareness and training for employees and contractors to ensure good security practices and prevent inadvertent or malicious intrusions into an organization's information systems.

Source: GAO analysis of NIST data.

In response to our recommendations, OMB has instructed federal agencies to ensure network security is incorporated into their agencywide network security program through the use of NIST guidelines. In addition, OMB's annual FISMA reporting requirements state that agencies must follow NIST standards and guidelines for non-national security programs and information systems. Since our report was issued, NIST has released guidelines that address the items identified in our recommendations. These guidelines include NIST SP 800-48, *Guide to Securing Legacy IEEE 802.11 Wireless Networks*; NIST SP 800-53, *Recommended Security Controls for Federal Information Systems and Organizations*; and NIST SP 800-97, *Establishing Wireless Robust Security Networks: A Guide to IEEE 802.11i* (see table 3).

Comprehensive Policies, Use of Secure Technologies, Risk-Based Approach, Training, and Monitoring Among Leading Practices for Deploying and Monitoring Secure Wireless Networks

Leading practices for deploying and monitoring secure wireless networks include comprehensive policies, configuration controls, training, and other practices as described in table 4. Many of these practices are consistent with the key information security controls required for an effective information security program identified in our previous reports and reflect wireless-specific aspects of those controls. Furthermore, experts identified several emerging technologies, such as broadband wireless, third-party device management, and IEEE 802.11n-2009/802.11w-2009, as potentially important in securing wireless networks in the future.

Table 4. Leading Practices for Securing Wireless Networks and Technologies

Practice category		Practice description
1.	**Policy**	Develop comprehensive security policies that govern the implementation and use of wireless networks and mobile devices that include the following safeguards: • implement secure encryption with enterprise authentication, • establish usage restrictions and implementation guidance for wireless access, and • enforce access controls for connection of mobile devices.
2.	**Risk-based approach**	Employ a risk-based approach for wireless deployment.
3.	**Centralized management**	Employ a centralized wireless management structure that is integrated with the existing wired network.
4.	**Configuration requirements**	Establish configuration requirements for wireless networks and devices in accordance with the developed security policies and requirements.
5.	**Training**	Incorporate wireless and mobile device security component in training.
6.	**Remote access**	Use a VPN to facilitate the secure transfer of sensitive data during remote access.
7.	**Monitoring**	Deploy continuous monitoring procedures for detecting rogue access points and clients using a risk-based approach.
8.	**Security assessments**	Perform regular security assessments to help ensure wireless networks are operating securely.

Source: GAO.

Develop Comprehensive Security Policies that Govern Implementation and Use of Wireless Networks and Mobile Devices

Comprehensive information security policies that address the security of wireless networks and mobile devices can help agencies mitigate risks. FISMA recognizes that the development of policies and procedures is essential to cost effectively reducing the risks associated with IT, including wireless IT, to an acceptable level. In addition, experts noted that sound policy is the basis for all effective security measures. Policies should cover areas such as roles and responsibilities, WLAN infrastructure security, WLAN client device security, and security assessments. By establishing policies that address these areas, agencies can create a framework for applying practices, tools, and training to help support the security of wireless networks.

In addition to these policies, federal guidelines and experts also emphasized key safeguards to address wireless security concerns that have surfaced since our 2005 report. These practices include prohibiting the use of WEP and implementing WPA2 with enterprise authentication, establishing usage restrictions and mplementation guidance for wireless access, and implementing access controls for mobile devices that connect to an agency's wireless networks.

Implement Secure Encryption with Enterprise Authentication

Organizations should establish policies requiring procurement of wireless products that have been WPA2-Enterprise certified and Federal Information Processing Standards (FIPS)-validated.[11] NIST guidelines state, and experts agree, that only these devices are capable of fully implementing the IEEE 802.11i robust security network protections, which include enhanced user and message authentication mechanisms, cryptographic key management, and robust enciphering and data integrity mechanisms. Wireless technologies that rely on older wireless security protocols, such as WEP and WPA, can be more easily exploited to circumvent or adversely impact access control and authentication, confidentiality, integrity, and availability since they do not require strong encryption algorithms.

Establish Usage Restrictions and Implementation Guidance for Wireless Access

Agencies should establish and enforce usage restrictions and implementation guidance for wireless access. According to NIST guidelines, security policies should identify which users are authorized to connect wirelessly to an organization's networks and the types of information allowed to be transmitted across wireless networks. In addition, wireless access to information systems should only be permitted by using authentication and encryption.

NIST guidelines also instruct agencies to identify the acceptable methods of remote access. Specifically, an agency's policies should describe which wireless-enabled devices can connect to the agency's networks remotely and the types of external networks permitted. For example, policies should specify if users connecting remotely through public hot spots to an agency's networks are authorized to use only agency-issued mobile devices.

Enforce Access Controls for Connection of Mobile Devices

Both NIST guidelines and experts identified establishing access controls for mobile devices, which includes those taken to locations the agency deems to be a significant risk, and prohibiting dual connection as a key practice for securely deploying and monitoring wireless networks. Our previous reports have also emphasized the importance of such access controls to limit, prevent, or detect inappropriate access to computer resources (data, equipment, and facilities), in order to protect them from unauthorized use, modification, disclosure, and loss.

Specifically, NIST guidelines state that agencies need to establish and enforce usage restrictions and implementation guidance for agency-issued mobile devices taken to locations the agency deems to be a significant risk. For example, agencies may issue specially configured mobile devices to individuals before traveling to risky locations, such as certain countries, which are in accordance with agency policies and procedures. Upon return from travel, agency-defined inspection and preventative measures can be applied to the mobile device such as re-imaging the hard disk drive on laptops. According to one expert's account, by developing access control requirements for global use of devices (i.e., managing devices that travel internationally), organizations can avoid compromises of devices or information that have occurred at organizations without these practices.

Another component of establishing access controls for mobile devices that NIST and experts identified was restricting the access rights of devices (and individuals) to an organization's network. Enforcing requirements such as usage restrictions, configuration management, and device identification and authentication, and disabling unnecessary hardware can prevent incidents of employees inadvertently connecting their devices to malicious entities and compromising the confidentiality, integrity, and availability of the organization's network.

Employ a Risk-Based Approach for Wireless Deployment

Federal guidelines underscore, and experts agree with, the need for a risk-based approach for deploying wireless networks and technologies. A risk-based approach attempts to ensure that risks to the organization are identified and prioritized so that available resources can be most effectively used in defending against the most significant threats, such as unauthorized access points or devices on the network, and to prevent the incurrence of undue risk.

Federal guidelines and experts also agreed that risks should be considered prior to acquisition of wireless technologies in order to implement management controls early on, determine which WLAN activities pose an acceptable risk, and identify the potential impact of the threat to the organization. The Committee on National Security Systems,[12] in its recently issued *Policy on Wireless Communications: Protecting National Security Information,* recommended that agencies consider protecting against the risks that can occur during various points of data transmission such as at the point of origin, when received, and while stored on wireless media.[13] It is only after considering and then managing such risks that organizations can make informed decisions such as whether or not to deploy wireless networks and technologies or determine the types of devices and extent of their usage throughout the organization. Moreover, since risks change over time, as a general practice, it is essential to periodically reassess risks and reconsider the appropriateness and effectiveness of the policies and controls organizations have selected to mitigate those risks.

Employ a Centralized Wireless Management Structure That Is Integrated with the Existing Wired Network

Security experts agreed that a centralized wireless management structure that is integrated with the management of the existing wired network can provide a more effective means to manage the wireless infrastructure and the information security program as a whole. A centralized structure can provide a coherent, consistent approach to managing the entire wireless network. For example, configuration changes can be centrally monitored, and with centralized reporting, management can have improved visibility and oversight of the organization's entire wireless network. Using tools that allow centralized management of WLANs can provide a management focal point and reduce the number of attack points in the network.

A centralized structure can also facilitate the development and implementation of standardized guidance, which allows organizations to consistently apply information security policies. Organizations can authorize the use of specific products, coordinate the installation of WLANs, and issue other such directives to provide a holistic approach for deploying and monitoring wireless activities. Such implementation can be centralized at the enterprise or component level, based on the business needs of the organization.

In conjunction with a centralized management structure, experts agreed the importance of integrating wireless management with management of the wired network. By extending established information security controls, such as intrusion detection systems and monitoring, from the wired to the wireless network, an organization is better positioned to both understand and defend its overall information security posture.

Experts agreed that a decentralized wireless management structure can result in disparate, ad hoc networks that operate and are managed separately. The existence of multiple networks that are independently managed can impede effective implementation and monitoring of security controls and inhibit sufficient oversight of the wireless network.

Although centralized wireless management can have many benefits, the level of centralization needs to be determined by business need and a risk-based approach. Centralization can be performed by agency components or departmentwide and should be balanced against the costs of centralization.

Existing federal guidelines recognize the benefits that centralized management of network services and information security can provide for federal agencies. For example, NIST guidelines state that centralized security management is an important consideration for managing mobile devices since it facilitates the configuration control and management processes that support compliance with an organization's security policy. Additionally, we previously reported that establishing a central management focal point for information security is essential to spotting trends, identifying problem areas, and determining whether policies and administrative issues are handled in a consistent manner.[14]

Establish Configuration Requirements for Wireless Networks and Devices in Accordance with the Developed Security Policies and Requirements

Establishing configuration requirements for wireless networks and devices can help ensure they are deployed in a secure manner in accordance with agency policies. For example, NIST SP 800-48 states that agencies should configure their wireless networks in accordance with established security policies and requirements. Establishing settings or configuration requirements for wireless access points can guide their placement and signal strength to minimize signal leakage and exposure to attacks.

In addition to access points, client devices should also be configured to enhance the wireless network security posture. According to NIST, securing the infrastructure without properly securing the client devices renders the entire wireless network insecure. Solutions such as enterprise servers can periodically communicate with managed mobile devices to ensure security and other configuration settings are correct and in compliance with policy.

Incorporate Wireless and Mobile Device Security Component in Training

NIST guidelines and experts stated that training employees and contractors in an organization's wireless policies is a fundamental part of ensuring that wireless networks are configured, operated, and used in a secure and appropriate manner. For security policies to be effective, those expected to comply with them must be aware of the policies. Additionally, FISMA mandates that agencies provide security awareness training for their personnel, including contractors and other users of information systems that support the operations and assets of the agency. NIST recommends that the security awareness training include the risks of wireless security and how to protect against those risks. In addition, NIST guidelines and experts agree that an agency's security training should include mobile device security that addresses (1) maintaining physical control over mobile devices, (2) protecting sensitive data on mobile devices with encryption, (3) disabling wireless interfaces on mobile devices when not needed, and (4) the procedures for reporting lost or stolen mobile devices. FISMA also requires agency chief information officers to ensure that personnel with significant information security responsibilities receive training with respect to their responsibilities.

Use a VPN to Facilitate the Secure Transfer of Data during Remote Access

A VPN can provide a secure communications mechanism for sensitive data transferred across multiple, public networks. For wireless technologies, VPNs are useful because they provide a way to secure WLANs that may be insecure, such as those at public hot spots, in homes, or other locations. According to NIST guidelines, federal agencies should consider using VPNs to protect the confidentiality of WLAN communications during remote access and telework and should configure the VPNs to use FIPS-140-2-validated encryption. Experts agreed on the use of VPNs as an integral security measure for an increasingly mobile workforce.

Deploy Continuous Monitoring Procedures for Detecting Rogue Access Points and Clients Using a Risk-Based Approach

Continuous monitoring is a means for an organization to ensure that security controls remain effective despite the planned and unplanned changes that can occur to an information system. OMB policy and NIST guidelines require agencies to implement a continuous monitoring approach for all information systems, including those using wireless technologies. According to NIST guidelines, agencies are required to monitor for unauthorized wireless access to information systems and should base their determination of the scope and frequency of such monitoring on an assessment of risk to the agency, the operational environment, the agency's requirements, and specific threat information. Continuous monitoring allows an organization to defend its security posture in a dynamic environment where threats, vulnerabilities, and technologies are constantly changing. Experts also noted the importance of continuously monitoring the wireless network for rogue access points and client devices. Documenting and implementing an approach to wireless monitoring that uses a risk-based approach helps to ensure that the scope and frequency of monitoring is appropriate for the threats facing the agency. As previously mentioned, centralized management tools can provide continuous monitoring capabilities for improved visibility and oversight of the organization's entire wireless network.

Both experts and NIST guidelines highlighted the importance of using a wireless intrusion detection system to continuously monitor an agency's wireless networks to detect and respond to malicious activities on the network before they inflict damage. These types of systems enable an organization's operations or security staff to determine whether unauthorized users or devices are attempting to access, have already accessed, or have compromised a WLAN. A wireless intrusion prevention system builds on the functionality of a wireless intrusion detection system by also automatically taking countermeasures against these unauthorized users or devices. These systems are able to monitor wireless data as it passes from wireless to wired networks. They can also detect misconfigured WLAN clients, rogue access points, ad hoc networks, and other possible violations of an organization's WLAN policy. In addition, these systems can position an organization to proactively assess its wireless network at regular intervals. However, a wireless intrusion detection or prevention system is a significant expense, and it may not be appropriate in all cases. For example, an agency may determine that a smaller agency location with lower risk systems may not warrant the expense that installing a wireless intrusion detection or prevention system may entail.

Other tools exist to detect rogue wireless client devices, such as handheld scanners and network authentication mechanisms, but these may not be as effective or easy to monitor as an intrusion detection system. Consistent with NIST guidelines, an organization should use a risk-based business case to determine the appropriate use of continuous monitoring solutions.

Perform Regular Security Assessments to Help Ensure Wireless Networks are Operating Securely

Experts and NIST guidelines both noted the importance of regular security assessments for checking the security posture of wireless networks and for determining corrective actions needed to ensure the wireless networks remain secure. Regular assessments help to determine whether wireless devices are transmitting correctly and are on the correct channels. Experts noted the importance of consistently and regularly performing security assessments in tandem with continuously monitoring the wireless network. In addition, organizations should maintain an inventory of access points deployed and their mobile devices to help identify rogue devices when conducting assessments. Assessments can help organizations to determine whether controls are appropriately designed and operating effectively to achieve the organization's control objectives.

Broadband Wireless, Device Management, and Newer WLAN Technologies Are Emerging Wireless Technologies

Several current and emerging technologies are important to consider for secure deployment of wireless technologies as follows:

- *Long-Term Evolution*—Long-Term Evolution is a fourth-generation wireless broadband technology that experts stated is expected to improve the speed and quality of service and provide scalable bandwidth capacity. It is also expected to improve security through enhanced encryption to prevent eavesdropping and user identity confidentiality to prevent tracking of specific users. One expert noted that most of the public safety broadband environments used for emergency communications at the state and local levels will adopt Long-Term Evolution and highlighted the importance of its effective implementation by government entities.
- *WiMAX*—Another form of broadband wireless technology known as WiMAX (Worldwide Interoperability for Microwave Access) is intended for wireless metropolitan area networking and is an effort to provide seamless mobile access in much the same way as wide-area cellular networks with higher transmission speeds. Security advantages of WiMAX include mutual device/user authentication, improved traffic encryption, and options for securing data within the core network.
- *Third-party device management*—The technological capabilities of a third-party vendor may provide a means for organizations to establish security for mobile devices. According to experts, a vendor that specializes in wireless security may be more up-to-date on security vulnerabilities and better equipped to assess the security of wireless

networks than an agency's own staff. Capabilities provided by a vendor can include incident management, triggers if a device is taken overseas, remote trouble shooting, and usage trends, among others.

- *IEEE 802.11n-2009/802.11w-2009 technologies*—Two additions to the 802.11 family of WLAN technologies–802.11n-2009 and 802.11w-2009–are expected to improve the performance and security of WLANs. The technologies specified in 802.1 1n-2009 increase WLAN speed, improve reliability, and extend the range of wireless transmissions. The 802.11w2009 encryption standard builds on the 802.1 1i framework to protect against certain types of attacks on WLANs.

Agencies Have Acted to Secure Wireless Networks, but Additional Steps Are Needed to Effectively Mitigate Security Challenges

Agencies have taken several steps to address the security of their wireless networks; however, these steps have not been fully and comprehensively applied across the government. Specifically, application was inconsistent among the agencies for most of the following leading practices:

- Most agencies developed policies that reflected NIST guidelines and leading practices, but gaps existed in these policies, particularly with respect to dual-connected laptops and use of mobile devices on international travel.
- All agencies required a risk-based approach for management of wireless technologies.
- Many agencies used a decentralized structure for management of wireless, limiting the potential standardization that centralized management can provide, and guidance on centralization is limited.
- The five agencies where we did detailed testing generally securely configured wireless access points, but they had numerous weaknesses in laptop and smartphone configurations. Gaps in governmentwide guidance on configuration contributed to these weaknesses.
- Most agencies were missing key elements related to wireless security in their security awareness training.

- Twenty agencies required encryption, and eight of these agencies specified that a VPN must be used during remote access; four agencies did not require encryption.
- Many agencies had insufficient practices for monitoring or conducting security assessments. Furthermore, federal guidance in this area lacks specificity.

Existing governmentwide guidance and oversight efforts do not fully address agency implementation of the leading practices. Until agencies fully address these practices, they will not have sufficient assurance that the risks to sensitive wireless systems, and sensitive data transmitted across or processed by those systems, are adequately safeguarded from inadvertent or deliberate misuse, fraudulent use, improper disclosure, or destruction. Also, until OMB and DHS ensure they have effective means for oversight of federal agencies' efforts to secure wireless networks they may lack full visibility of the vulnerability of these networks to attack.

Agencies Have Developed Policies to Support Secure Use of Wireless Technologies, but Gaps Exist

Almost all agencies required wireless networks to employ encryption, but not all agencies required secure forms of encryption. Specifically, 23 of 24 agencies specified in their policies that agency wireless networks are required to employ encryption. However, 7 of the 23 agencies did not require secure forms of encryption. Specifically, 2 agencies' policies required the use of WEP, an older wireless encryption method that is vulnerable to attack; one agency required the use of WPA, which is not compliant with federal requirements for encryption; and 4 other agencies did not specify any type of encryption or require the use of FIPS 140-2 compliant encryption. In addition, 1 agency did not have any documented requirements for wireless transmissions to be encrypted, even though that agency has a wireless network deployed at its headquarters.

In certain cases, agency policies had been developed several years ago. Agencies had also not always updated their policies to reflect their implementations of WLANs or federal requirements for wireless encryption. Agencies that do not require the use of strong, FIPS-validated encryption algorithms on their wireless networks have limited assurance that sensitive agency information is being adequately protected from unauthorized disclosure or modification.

Most Agencies Have Established Usage Restrictions and Implementation Guidance for Wireless Networks

Twenty-three of the 24 agencies provided specific guidance to agency personnel on the types of information that may be transmitted using wireless networks or on how sensitive information is to be protected when transmitted wirelessly. All 24 of the agencies in our review had also developed policies establishing usage restrictions and implementation guidance for wireless networks; although policies for 3 agencies were in draft and had not yet been approved.

Examples of usage restrictions in agency policies included the following:

- requiring that administration of wireless infrastructure devices (such as access points) be conducted using the wired network,
- prohibiting the use of ad hoc wireless networks, and
- allowing access to agency wireless networks only via a VPN.

Agencies' policies frequently contained wireless implementation guidance such as the following:

- physically securing wireless infrastructure devices,
- adjusting the transmission power of access points to ensure adequate coverage while minimizing signal leakage,
- maintaining audit logs on wireless access points,
- changing default service set identifiers[15] and not using identifiers that would identify the agency,
- enabling media access control[16] address filtering, and
- segregating wireless network traffic from the wired network using firewalls or other methods.

Agencies Established Policies for Access Controls for Mobile Devices, but Several Agencies Did Not Specifically Address Wireless Functionality of Laptops, Dual Connection of Laptops, or International Travel of Mobile Devices

Almost all agencies had established some type of access control policy for mobile devices. Specifically, all 24 agencies developed policies for PDAs, such as smartphones, although 2 agencies' policies had not been finalized. Although 23 of the 24 agencies had developed implementation guidance for laptop computers, the policies of 4 of these agencies did not specifically address wireless functionality on laptops. In addition, 1 of the 24 agencies did not document laptop policies at all.

Fewer agencies had developed access control policies regarding dual connection of laptops. NIST guidelines recommend that client devices, such as laptop computers, should be configured not to allow the simultaneous use of more than one network interface. Although most agencies had established policies for wireless-enabled laptops, many did not address the risk of dual connections of laptops in their policies. As described earlier, the security of an agency network could be compromised when a laptop is connected to an external wireless network, and to an agency's wired network simultaneously, leaving it vulnerable to attack and providing unauthorized access to the wired network. Turning off or disabling the wireless capability when a laptop is connected to a wired network mitigates this risk. Of the 24 agencies in our review, 8 did not have documented policies requiring the wireless capability to be turned off or disabled when the agency's laptop is connected to a wired network. One agency with a decentralized wireless management structure had a high level overall wireless policy, but allowed its components to determine whether to augment it with more detailed policies, including policies prohibiting multiple network connections. Other agency officials incorrectly thought that the dual connection issue was addressed by governmentwide guidance such as the Federal Desktop Core Configuration (FDCC).[17]

Although the baseline FDCC standard disables wireless connectivity, in March 2010,[18] we reported that many agencies have chosen to deviate from the standard and enable wireless functionality on their workstations. No other setting or combination of settings within the FDCC standard prevents multiple network connections. We also previously reported that OMB had not specified any guidance for agencies to use when considering the risks of deviating from the FDCC standard; OMB has therefore not specified any such guidance for agencies regarding permitting the use of wireless technologies. We recommended that OMB clarify its policies regarding FDCC deviations to include guidance for agencies to use when assessing the risks of deviations. Until OMB provides guidance to agencies regarding the risks associated with enabling wireless on agency laptops, including the risk of dually connected laptops, agencies may not document and implement policies prohibiting dual connections, increasing the risk that an attacker would be able to gain unauthorized access into an agency's network and destroy, modify, or copy sensitive information. Further, until agencies fully document and implement policies prohibiting dual connections, an increased risk exists that an attacker would be able to gain unauthorized access into an agency's network and destroy, modify, or copy sensitive information.

Similarly, many agencies also did not have documented policies governing international travel with mobile devices. As noted earlier, according to NIST, a leading practice for client and mobile devices is to issue specially configured laptops, PDAs, and other mobile devices to individuals traveling to locations considered to be high risk and to apply preventative measures to devices being returned from such locations. However, only 12 of the 24 agencies had documented policies for safeguarding PDAs taken internationally, and policies for 4 agencies were in draft. Policies of 5 agencies required specially configured devices to be issued for such travel, although policies of 10 agencies required preventative measures to be applied to the devices after they were returned from travel and before being connected to agency networks. In addition, just 9 of the 24 agencies had documented policies for laptops taken internationally; including 2 agencies that had draft policies. Only 4 of the 9 agencies required that a specially configured laptop be used for travel, although 8 agencies required preventative measures to be applied after the devices were returned from travel.

Several agency officials stated that they were aware of the risks posed to mobile devices during international travel, but that agencies had not yet developed policies to address these risks. NIST issued its updated guidelines on this practice in August 2009, and many agencies had not yet updated their security policies to reflect the new guidelines. By not having documented policies, agencies may be at increased risk that sensitive information could be compromised while a device is in another country, or that malware obtained during an international trip could be inadvertently introduced onto agency networks, placing sensitive data and systems at risk.

All Agencies Required a Risk-Based Approach to Wireless Deployment, and Most Required Approval of New Wireless Technologies

All of the agencies in our review required in their policies that decisions related to management of wireless technologies be based on risk. Fifteen agencies had policies that specifically required a risk-based approach to wireless management; the remaining 9 agencies had policies that, while not specific to wireless, required a risk-based approach to all management of IT.

Twenty-two of the 24 agencies had documented policies specifying that new wireless technologies require approval from an appropriate official or governing body before they could be implemented, although one agency's policy was still in draft. The remaining 2 agencies, while not specifying wireless, required all new technologies to be approved.

Many Agencies Did Not Use a Centralized Structure for Wireless Management

Although a centralized wireless management structure can provide a more effective means of managing wireless networks, many agencies reported not using a centralized approach. Eleven agencies indicated that they did have a centralized wireless management structure, and 3 indicated using both a centralized and decentralized structure, depending on agency component. Ten agencies employed a decentralized wireless management structure. As previously mentioned, a decentralized wireless management structure can result in disparate, ad hoc networks that are independently managed, which can impede effective implementation and monitoring of security controls and inhibit sufficient oversight of the wireless network. The following examples describe 2 agencies that are implementing a centralized management approach and identify the benefits and limitations of their implementation efforts.

One agency where we performed detailed testing had deployed a centrally monitored and managed wireless intrusion detection system that was integrated with the agency's national wired networks and operated centrally by its cybersecurity management center. This system had several positive aspects such as eliminating the need for trained personnel in every location, easy integration with other automated tools, and a console that provided a summary view of security events and detected devices. Additionally, it provided a central means to monitor configurations of wireless devices, discover rogue access points, and detect intrusion attempts. However, this system also had limitations. According to agency officials, the center did not always manage the installation and configurations of wireless devices at the facilities being monitored. As a result, the agency could not take full advantage of the improved visibility and oversight of the agency's entire wireless network that centralized management can offer.

Another agency was deploying a centrally managed WLAN nationwide to hundreds of facilities. According to agency officials, the WLAN would provide a platform for numerous systems and devices to be operated in a highly mobile environment. The centrally managed WLAN has the potential to simplify and provide more control over WLAN management. For instance, configuration policies created in templates can be forwarded to all controllers connected to the network and then on to the wireless access points. It can also provide a graphical display of the WLAN and its performance.

Agencies had decentralized approaches to wireless management for several reasons. Several agencies managed IT in a decentralized manner, delegating responsibility to agency components based on their business needs. Other agencies were just beginning to consider use of WLANs, or had only deployed WLANs in a limited manner. Technological advances have also made centralized management of wireless more feasible than in the past. Furthermore, while existing federal guidelines recognizes the benefits that centralized management of information security can provide for federal agencies; existing NIST guidelines on wireless security do not provide detail on the appropriate ways agencies can centralize their management of wireless technologies based on business need.

Until agencies effectively implement a centralized wireless management structure, they will have limited visibility and control over WLANs and a limited ability to integrate wireless management controls with the existing wired network to provide continuity and robustness to their organization's overall information security program.

Although Access Points Were Generally Configured Securely, Weaknesses Existed in Configurations of Laptops, Firewalls, and Smartphones Used for Wireless Access

At the five agencies where we conducted detailed testing, the access points used to provide WLANs were generally configured in a secure manner. However, we identified weaknesses in laptop or firewall configurations at these agencies as the following examples illustrate:

- None of the five agencies had fully implemented controls to prevent laptops from connecting to a wireless network while also being connected to the agency's internal wired network. As described earlier, when a laptop is connected to a wireless network and to an agency's wired network simultaneously, an attacker could exploit this dual connection to gain unauthorized access to the agency's network, placing sensitive information and systems at risk. In certain cases, agency officials were unaware that the potential for a dual connection existed. In other cases, officials were aware of the risk, but were unsure of the appropriate controls that could mitigate this risk, or were concerned that these controls might interfere with needed functionality of the device. In general, workstations using Microsoft Windows require an additional third-party application or other specialized configuration to disable wireless connectivity. Although NIST guidelines recommend that client devices, such as laptop computers, be configured to not allow the simultaneous use of more than one network interface, existing NIST guidelines on wireless networks do not provide specific technical information on steps that agencies should take to implement this control.

- One of the five agencies had configured a laptop to allow nonprivileged users to enable Bluetooth and to connect to other personal Bluetooth devices. Furthermore, Bluetooth was configured to default to "discovery" mode, making the laptops visible to other Bluetooth devices. As a result, an attacker with a Bluetooth device within range could connect to the agency's laptop, providing a means of unauthorized access to sensitive information on the laptop itself, as well as to the agency's network.
- Two agencies allowed general users to have administrative privileges on their laptops, thus reducing the effectiveness of established security controls on those machines and increasing the risk that users could install unapproved and potentially malicious software, which could allow sensitive information to be viewed, modified, or deleted.
- At one agency, a firewall[19] segmented a guest wireless network from the agency's internal network. However, the firewall was configured to allow all traffic that used Internet protocol version 6 (IPv6) to flow between the networks without controls. As a result, any user—whether malicious or not—connected to the guest wireless network using the IPv6 protocol could traverse the guest network without any authentication or access controls and could potentially gain unauthorized access to the internal network, placing sensitive agency information and systems at risk of unauthorized disclosure, modification, misuse, or destruction.

Many agencies also did not enforce secure configurations on their BlackBerry smartphones. DISA has developed a configuration checklist to help its administrators securely configure its BlackBerry Enterprise Servers, which are servers that allow agencies to centrally control security policy for BlackBerry smartphones. These guidelines have also been made available to other organizations, including federal agencies, as part of a NIST program providing secure configurations for computing devices. Although not mandatory, the guidelines provide a starting point for securely configuring BlackBerry smartphones. However, 18 of the 24 agencies had server configurations that were less secure than the DISA guidelines. For example:

- Fourteen agencies allowed BlackBerry passwords of insufficient length. DISA recommends that passwords on BlackBerry smartphones be a minimum of eight characters in length.
- Seven agencies did not require the use of complex passwords. DISA recommends that this value be configured to require passwords to contain, at a minimum, at least one alphabetic character and one numeric character.
- Eleven agencies did not set a sufficient security timeout period. DISA recommends that this value be set to 15 minutes or less.
- Ten agencies did not configure a setting that prevents applications from opening internal and external connections simultaneously, exposing the device to malware.

Several agency officials stated that the DISA checklist was not mandatory for federal agencies; however, no other federal configuration standard for BlackBerry smartphones currently exists. Due to their portability and capacity to collect and store significant amounts of sensitive information, smartphones such as the BlackBerry are susceptible to security threats such as loss, theft, unauthorized access, malware, electronic eavesdropping, and tracking. Without securely configuring their BlackBerry Enterprise Servers, agencies are at an increased risk that their BlackBerry smartphones could be compromised, resulting in tampered, lost, or stolen data.

Agencies Have Improved Wireless Security Training Efforts, but Training Often Lacked Key Elements

Many agencies did include key information on the risks of wireless technologies and how to mitigate such risks in their training programs. Specifically, 18 agencies provided training on the inherent lack of security of wireless technology and gave information on how employees and contractors could protect information that is transmitted wirelessly. However, 6 agencies did not address wireless security in their annual training.

In addition, although most agencies included information on mobile devices in their security awareness training, most agencies did not include key elements in accordance with NIST guidelines. Specifically, only 2 of the 24 agencies included in their training that users should disable the wireless interfaces on their mobile devices when not needed. In addition, training at 14 agencies did not address physical control over mobile devices; 5 did not describe the procedures for reporting lost or stolen mobile devices; and 5 did not include information on encrypting sensitive data on mobile devices. Finally, 1 agency did not address mobile device security in its annual training.

Awareness about wireless security challenges can assist employees in complying with policies and procedures to reduce agency information security risks. Without such training, employees and contractors may practice behaviors that threaten the safety of the agency's data.

Agency Policies Did Not Always Require the Use of a VPN or Encryption for Remote Access

Policies on remote access are important to the security of wireless devices because a frequent use of wireless technologies is for access to agency networks from remote locations, such as a home or hotel WLAN. Twenty of the 24 agencies required remote access sessions to be encrypted, and 8 of these agencies specified that a VPN must be used. However, 4 of the 24 agencies did not require remote access to be encrypted using a VPN or other encryption method. Without having policies requiring remote access sessions to employ adequate encryption, agencies will not be able to ensure that sensitive information is protected from unauthorized access, use, disclosure, or modification when users connect to agency information systems remotely.

Many Agency Policies and Practices for Monitoring 802.11 Networks and Conducting Assessments Were Insufficient

All 24 agencies in our review reported some form of monitoring for the existence of unauthorized or "rogue" wireless networks. Sixteen agencies reported that they continuously monitored 24 hours a day at one or more agency facilities. However, we found significant weaknesses in agency policies for wireless monitoring. Only 18 agencies required any type of monitoring for unauthorized access points in their policies, sometimes as rarely as once per year. In addition, two agencies used outdated scanning tools that could miss key wireless activities. Six agencies lacked any requirements for wireless monitoring. This lack of requirements, combined with the ease of setting up wireless networks, creates a situation in which wireless networks can be operating in these agencies without authorization or the required security configurations.

At the five agencies where we performed detailed testing, we found that the approach that several locations took toward monitoring and assessments for 802.11 wireless activity had significant weaknesses. Five agency locations did not have routine procedures for performing wireless assessments for unauthorized devices and networks. Two of these locations had not performed wireless scans in the past 2 years; two other agency locations did not document the results of scans.

One agency where we performed detailed testing had deployed a centrally monitored and managed wireless intrusion detection system at one of its locations. However, according to agency officials, because of the costs of the system, it was not deployed to all locations. At the location we visited that did not have the system deployed, there was no alternate approach to wireless monitoring, posing the risk of undetected wireless access points, intrusions, and loss of sensitive, proprietary data.

Further, while three other agencies also used a wireless intrusion detection system at some locations to continuously monitor for unauthorized devices and networks, the monitoring at these locations was ineffective. Specifically, the systems at each location had not been tailored to ignore known false positives. As a result, the systems generated large numbers of alerts for rogue access points, most of which were false. Local network administrators therefore had no way to determine which alerts were actual security events, hindering their ability to take advantage of the security aspects of the system.

Although NIST guidelines recommend that agencies use wireless monitoring, it does not specify criteria for selecting tools to ensure they provide comprehensive monitoring capabilities, nor does it suggest appropriate frequencies for recurring assessments or recommendations for when continuous monitoring may be appropriate.

Regular monitoring and security assessments are key practices for ensuring the security of wireless networks and devices. Even at agencies that have no wireless networks deployed, wireless-enabled devices that are deployed on the network, such as laptop computers, can provide a potential means for an attacker to gain unauthorized access to the network, putting critical agency systems and information at risk of unauthorized modification, misuse, disclosure, or destruction. Until regular monitoring and assessment policies and practices are implemented, these networks are at increased vulnerability to attack.

Existing Governmentwide Reporting and Oversight Efforts Do Not Fully Address Key Wireless Security Practices

The annual FISMA reporting process administered by OMB (and recently devolved from OMB to DHS by an OMB memorandum), which serves as a means of oversight of federal agency information security, does not fully address implementation of leading practices in wireless security. As of October 2010, the fiscal year 2010 draft reporting metrics do contain measures related to automated configuration management, vulnerability management, and incident management. However, they do not include specific metrics related to wireless security issues identified in this report, such as measures to address the risk of dual-connected laptops, policies related to international travel with mobile devices, the extent to which agencies have centralized their management of wireless devices, and agency practices for monitoring and assessment of wireless networks.

Furthermore, although the DHS official responsible for the agency's newly assigned governmentwide FISMA compliance activities stated that the agency plans additional activities that may address aspects of wireless security governmentwide, the scope and time frames for these activities have not yet been finalized.

Until OMB and DHS ensure they have effective means for oversight of federal agencies' efforts to secure wireless networks, they lack full visibility of the vulnerability of these networks to attack.

CONCLUSIONS

Federal agencies are making significant use of wireless networks and devices, including WLANs, laptop computers, and smartphones. Several leading practices exist to secure these technologies, including developing comprehensive policies, employing a centralized approach to management, establishing secure network and device configurations, and having effective training and monitoring in place.

Agencies have taken several steps to address the security of their wireless networks and devices, including development of security policies, centralized management, training, and monitoring; however, these steps have not been fully and comprehensively applied across the government. Gaps exist in policies, network management was not always centralized, and numerous weaknesses existed in configurations of laptops and smartphones. Particular issues are the risk of dual-connected laptops and risks related to mobile devices being taken on international travel In addition, many agencies had insufficient policies and practices for monitoring or conducting assessments of wireless technologies. Until OMB, DHS, NIST, and individual agencies take steps to fully implement leading security practices, federal wireless networks will remain at increased vulnerability to attack, and information on these networks is subject to unauthorized access, use, disclosure, or modification.

RECOMMENDATIONS FOR EXECUTIVE ACTION

To improve governmentwide oversight of wireless security practices, we recommend that the Director of OMB, in consultation with the Secretary of Homeland Security, implement the following two

- include metrics related to wireless security as part of the FISMA reporting process, and
- develop the scope and specific time frames for additional activities that address wireless security as part of their reviews of agency cybersecurity programs.

We also recommend that the Secretary of Commerce instruct the Director of NIST to develop and issue guidelines in the following four areas:

- technical steps agencies can take to mitigate the risk of dual connected laptops,
- governmentwide secure configurations for wireless functionality on laptops and for smartphones such as BlackBerries,
- appropriate ways agencies can centralize their management of wireless technologies based on business need, and
- criteria for selection of tools and recommendations on appropriate frequencies of wireless security assessments and recommendations for when continuous monitoring of wireless networks may be appropriate.

In addition, in a separate report with limited distribution, we are making 134 recommendations to 24 major federal agencies to address weaknesses in wireless-related information security controls, including policies, procedures, and technical configurations.

AGENCY COMMENTS AND OUR EVALUATION

We provided a draft of this report to the Director of OMB and the Secretary of Commerce for their review and comment. However, OMB did not provide comments on the report.

In written comments on a draft of this report, the Secretary of Commerce stated that the department concurred with our recommendations that NIST develop additional guidance related to wireless security. The Secretary also suggested that we use the term "NIST guidelines" rather than "NIST guidance" throughout the report, in addition to other technical comments. We have incorporated these comments in the report where appropriate.

Gregory C. Wilshusen
Director, Information Security Issues

Dr. Nabajyoti Barkakati
Chief Technologist

APPENDIX I. OBJECTIVES, SCOPE, AND METHODOLOGY

The objectives of our review were to (1) identify leading practices and state-of-the-art technologies for deploying and monitoring secure wireless networks and (2) assess agency efforts to secure wireless networks, including vulnerability to attack. The scope of our review included the 24 major federal agencies covered by the Chief Financial Officers Act.[20]

To identify leading practices for deploying and monitoring secure wireless networks, we first identified subject matter experts, including leading organizations and individuals, by reviewing information security-related Web sites and professional literature. In addition, we identified organizations that received recognition based on an industry magazine's rankings for top wireless or other information security-related products. We also solicited suggestions on subject matter experts from individuals working in the field of wireless security at major information technology (IT) and telecommunications companies and federal government agencies, such as the National Institute of Standards and Technology (NIST), National Security Agency, and Committee on National Security Systems, because they were in a position to evaluate and compare wireless security practices at numerous organizations. We contacted approximately 10 organizations and individuals that met the above criteria; 8, including 5 organizations and three individuals, agreed to be interviewed and provide input on the practices we identified. The organizations were prominent and nationally known, and the individuals were recognized as experts in the information security community. The participants included a wireless services provider, a global technology products and services provider, a global telecommunications provider, a nonprofit industry organization, a standards laboratory, a government information security consortium, a defense agency, and a wireless security consultant.

Then, to determine the specific leading practices, we obtained information, primarily through analysis of publications, guidance, checklists, presentations, and other documentation, and interviews with subject matter experts. We supplemented the information gathered with information obtained from our professional literature review. We then analyzed the information obtained to identify common wireless security leading practices and validated the practices we identified with the subject matter experts.

To assess agency efforts to secure wireless networks, we obtained and analyzed documents such as departmental and component policies, plans, configuration documents, and training materials to determine the extent of wireless technologies used and the security controls implemented at each of the 24 major federal agencies. We also obtained information through structured interviews with officials responsible for wireless security policies and practices for each of the 24 agencies. For each of these agencies, we used a laptop equipped with an antenna that served as a mobile scanning device and walked or drove around the perimeters of publicly accessible areas of their headquarters facilities in the Washington, D.C., area to collect data to determine wireless technologies that were deployed in the buildings. We also conducted scans of multiple agency facilities in another major metropolitan area outside of the Washington, D.C., region. This area was chosen based on the following criteria:

- contained regional offices for the multiple major federal agencies in locations that were sufficiently dispersed to not have too many 802.11 signals within a narrow proximity; and
- had several regional offices with additional field offices nearby.

Based on the initial data collected from scans at the headquarters and field locations, we chose 5 of the 24 agencies at which to complete additional detailed wireless security testing, specifically, the Departments of Agriculture, Commerce, Transportation, and Veterans Affairs, and the Social Security Administration. These agencies were selected based on several criteria, including the amount of usage of wireless technologies, the level of centralization of IT management, and potential security issues revealed by the initial scan results. More in-depth testing at these agencies included a review of the configurations of client devices, wireless infrastructure, and monitoring practices. We inspected client devices to determine if security controls had been implemented to protect the local network. We also examined each agency's network infrastructure to determine if access points were encrypted and configured to deny unauthorized access. Finally, we determined if the agencies monitored the IEEE 802.11 wireless spectrum.

We conducted this performance audit from January 2010 to November 2010, in accordance with generally accepted government auditing standards. Those standards require that we plan and perform the audit to obtain sufficient, appropriate evidence to provide a reasonable basis for our findings and conclusions based on our audit objectives. We believe that the evidence obtained provides a reasonable basis for our findings and conclusions based on our audit objectives.

End Notes

[1] GAO, *Information Security: Federal Agencies Need to Improve Controls over Wireless Networks*, GAO-05-383 (Washington, D.C.: May 17, 2005).

[2] H.R. Conf. Rep. No. 111-366, at 914 (2009). We briefed the committees on the preliminary results of our review on April 13, 2010.

[3] The 24 major federal agencies are the Agency for International Development; the Departments of Agriculture, Commerce, Defense, Education, Energy, Health and Human Services, Homeland Security, Housing and Urban Development, the Interior, Justice, Labor, State, Transportation, the Treasury, and Veterans Affairs; the Environmental Protection Agency; the General Services Administration; the National Aeronautics and Space Administration; the National Science Foundation; the Nuclear Regulatory Commission; the Office of Personnel Management; the Small Business Administration; and the Social Security Administration.

[4] IEEE is a professional association focused on electrical and computer sciences, engineering, and related disciplines. IEEE is responsible for developing technical standards through the IEEE Standards Association, which follows consensus-based standards development processes.

[5] The Wi-Fi Alliance is a nonprofit international association that has the goal of certifying the interoperability of WLAN products based on IEEE 802.11 specifications.

[6] The Bluetooth Special Interest Group is a not-for-profit trade association developed to serve as the governing body for Bluetooth specifications.

[7] A VPN is a private network that is maintained across a shared or public network, such as the Internet, by means of specialized security procedures. VPNs are intended to provide secure connections between remote clients, such as branch offices or traveling personnel and a central office.

[8] 44 U.S.C. § 3544(b).

[9] OMB, *Clarifying Cybersecurity Responsibilities and Activities of the Executive Office of the President and the Department of Homeland Security* (Washington, D.C: July 6, 2010).

[10] GAO-05-383.

[11] See NIST, *Security Requirements for Cryptographic Modules*, FIPS 140-2 (Gaithersburg, MD: May 25, 2001). FIPS 140-2 specifies the security requirements for a cryptographic module used within a security system protecting sensitive information in computer and telecommunication systems (including voice systems) and provides four increasing, qualitative levels of security intended to cover a wide range of potential applications and environments. Agencies are required to encrypt agency data, where appropriate, using NIST-validated cryptographic modules as specified in FIPS 140-2

[12] The Committee on National Security Systems consists of 21 members from federal agencies and is charged with establishing national policy and promulgating direction, operational procedures, and guidance for the security of national security systems.

[13] Committee on National Security Systems, *Policy on Wireless Communications: Protecting National Security Information*, CNSSP No.17 (Ft. Meade, MD: May 2010).

[14] See, for example, GAO, *Executive Guide: Information Security Management, Learning From Leading Organizations*, GAO/AIMD-98-68 (Washington, D.C.: May 1998).

[15] A service set identifier is a name assigned to a wireless network that allows wireless clients to distinguish one wireless network from another.

[16] The media access control address is a unique identifier assigned to network adapters usually by the manufacturer for identification. Although intended to be a permanent and unique identification, it is possible to change the media access control address on most hardware.

[17] The FDCC was an initiative launched by OMB to require federal agencies to implement common security configurations on Microsoft Windows XP and Vista operating systems. The initiative has evolved into the United States Government Configuration Baseline, run by NIST.

[18] GAO, *Information Security: Agencies Need to Implement Federal Desktop Core Configuration Requirements*, GAO-10-202 (Washington, D.C.: Mar. 12, 2010).

[19] A firewall is a hardware or software component that protects given computers or networks from attacks by blocking network traffic or by allowing only authorized protocols and services to cross the boundary between networks.

[20] The 24 major federal agencies are the Agency for International Development; the Departments of Agriculture, Commerce, Defense, Education, Energy, Health and Human Services, Homeland Security, Housing and Urban Development, the Interior, Justice, Labor, State, Transportation, the Treasury, and Veterans Affairs; the Environmental Protection Agency; the General Services Administration; the National Aeronautics and Space Administration; the National Science Foundation; the Nuclear Regulatory Commission; the Office of Personnel Management; the Small Business Administration; and the Social Security Administration.

In: Federal Information Security ISBN 978-1-61942-771-6
Editors: D. G. Miller and I. I. Wilson

Chapter 2

INFORMATION SECURITY: FEDERAL DEPOSIT INSURANCE CORPORATION NEEDS TO MITIGATE CONTROL WEAKNESSES

United States Government Accountability Office

WHY GAO DID THIS STUDY

The Federal Deposit Insurance Corporation (FDIC) has a demanding responsibility enforcing banking laws, regulating financial institutions, and protecting depositors. Because of the importance of its work, the corporation must employ strong information security controls to ensure that its information systems are adequately protected from inadvertent misuse, fraud, and improper disclosure.

As part of its audit of the 2009 financial statements of the Deposit Insurance Fund and the Federal Savings & Loan Insurance Corporation Resolution Fund administrated by FDIC, GAO assessed (1) the effectiveness of FDIC's controls in protecting the confidentiality, integrity, and availability of its financial systems and information and (2) the progress FDIC has made in mitigating previously reported information security weaknesses. To perform the audit, GAO examined security policies, procedures, reports, and other documents; tested controls over key financial applications; and interviewed key FDIC personnel.

What GAO Recommends

GAO is recommending that FDIC improve key information activities to enhance the corporation's information security program. FDIC generally agreed with GAO's recommendations and stated that it plans to address the identified weaknesses.

What GAO Found

FDIC did not sufficiently implement access and other controls intended to protect the confidentiality, integrity, and availability of its financial systems and information. For example, it did not always

- sufficiently restrict user access to systems,
- ensure strong system boundaries,
- consistently enforce strong controls for identifying and authenticating users,
- encrypt sensitive information, or
- audit and monitor security-relevant events.

In addition, FDIC did not have policies, procedures, and controls in place to ensure the appropriate segregation of incompatible duties, adequately manage the configuration of its financial information systems, and update contingency plans. A key reason for these weaknesses is that FDIC did not always fully implement key information security program activities such as effectively developing, documenting, and implementing security policies, and implementing an effective continuous monitoring program. Until these weaknesses and program deficiencies are corrected, the corporation will not have sufficient assurance that its financial information and assets are adequately safeguarded from inadvertent or deliberate misuse, fraudulent use, improper disclosure, or destruction.

Despite the newly identified weaknesses, FDIC has mitigated each of the information security weaknesses previously reported by GAO. To its credit, the corporation has made improvements to its configuration management controls and aspects of its security management. For example, it maintained a full and complete requirements baseline for two systems and included key information in a remedial action plan.

Nevertheless, GAO concluded that weaknesses in information security controls constituted a significant deficiency in internal controls over the information systems and data used for financial reporting. Until FDIC corrects the security weaknesses identified during this year's audit, it will face an elevated risk of the misuse of federal assets, unauthorized modification or destruction of financial information, inappropriate disclosure of other sensitive information, and disruption of critical operations.

ABBREVIATIONS

DHS	Department of Homeland Security
FDIC	Federal Deposit Insurance Corporation
FISMA	Federal Information Security Act
ID	identification
IP	Internet Protocol
NIST	National Institute of Standards and Technology
NSA	National Security Agency
OMB	Office of Management and Budget
SNMP	Simple Network Management Protocol
US-CERT	United States Computer Emergency Readiness Team

November 30, 2010
The Honorable Sheila C. Bair
Chairman
Federal Deposit Insurance Corporation

Dear Madame Chairman:

The Federal Deposit Insurance Corporation (FDIC) has a demanding responsibility enforcing banking laws, regulating banking institutions, and protecting depositors. In carrying out its financial and mission-related operations, FDIC relies extensively on computerized systems. Because FDIC plays an important role in maintaining public confidence in the nation's financial system, issues that affect the confidentiality, integrity, and availability of the sensitive information maintained on its systems are of paramount concern. In particular, effective information security controls are essential to ensure that FDIC systems and information are adequately protected from inadvertent or deliberate misuse, fraudulent use, improper disclosure, or destruction.[1]

As part of our audit of FDIC's calendar year 2009 financial statements of the Deposit Insurance Fund and the Federal Savings & Loan Insurance Corporation Resolution Fund, we assessed the effectiveness of FDIC's information security controls over key financial systems, data, and networks.[2] In that report, we concluded that weaknesses in information security controls collectively constituted a significant deficiency in internal controls over the information systems and data used for financial reporting.[3]

In this report, we provide additional details on FDIC's information security controls during calendar year 2009. Our specific objectives were to assess (1) the effectiveness of its controls for ensuring the confidentiality, integrity, and availability of its financial information systems and information and (2) the status of FDIC's actions to correct or mitigate previously reported information security weaknesses. We conducted this performance audit at FDIC facilities in Arlington, Virginia; Washington, D.C.; and Dallas, Texas, from December 2009 to November 2010 in accordance with generally accepted government auditing standards.[4] Those standards require that we plan and perform the audit to obtain sufficient, appropriate evidence to provide a reasonable basis for our findings and conclusions based on our audit objectives. We believe our audit provides a reasonable basis for our findings and conclusions. See appendix I for additional details on our objectives, scope, and methodology.

BACKGROUND

Information security is a critical consideration for any organization that depends on information systems and computer networks to carry out its mission and is especially important for a government corporation such as FDIC, which oversees the financial institutions that are entrusted with safeguarding the public's money. While the use of interconnected electronic information systems allows FDIC to accomplish its mission more quickly and effectively, their use also exposes FDIC's information to the various internal and external threats that come with the use of such systems.

Cyber-based threats to information systems and cyber-related critical infrastructure are evolving and growing and can affect FDIC information systems and computer networks. Threats that could adversely affect the systems and information relevant to FDIC's operations associated with financial management and reporting can come from sources internal and external to the organization. Internal threats include mistakes by individuals, and fraudulent or malevolent acts by insiders. External threats include the ever-growing number of cyber-based attacks that can come from a variety of sources such as hackers, criminals, and foreign nations.

These potential attackers have a variety of techniques at their disposal, which can vastly enhance the reach and impact of their actions. For example, cyber attackers do not need to be physically close to their targets, their attacks can easily cross state and national borders, and cyber attackers can readily preserve their anonymity. Further, the interconnectivity among information systems presents increasing opportunities for such attacks. Indeed, reports of security incidents from federal agencies are on the rise, increasing by more than 400 percent from fiscal year 2006 to fiscal year 2009. Specifically, the number of incidents reported by federal agencies to the United States Computer Emergency Readiness Team (US-CERT) has increased dramatically over the past 4 years: from 5,503 incidents reported in fiscal year 2006 to about 30,000 incidents in fiscal year 2009.[5]

Compounding the growing number and kinds of threats are the significant deficiencies in security controls on the information systems at federal agencies, which have resulted in vulnerabilities in both financial and nonfinancial systems and information. These deficiencies continue to place assets at risk of inadvertent or deliberate misuse, financial information at risk of unauthorized modification or destruction, and critical operations at risk of disruption.

Accordingly, we have designated information security as a governmentwide high-risk area since 1997, a designation that remains in force today.[6] Recognizing the importance of securing federal agencies' information systems, Congress enacted the Federal Information Security Management Act (FISMA) in December 2002 to strengthen the security of information and systems within federal agencies.[7] FISMA requires each agency to develop, document, and implement an agencywide information security program to provide information security for the information and systems that support the operations and assets of the entities, using a risk- based approach to information security management.

FDIC Is a Key Protector of Bank and Thrift Deposits

FDIC was created by Congress to maintain the stability of and public confidence in the nation's financial system by insuring deposits, examining and supervising financial institutions, and resolving troubled institutions. Congress created FDIC in 1933[8] in response to the thousands of bank failures that occurred in the 1920s and early 1930s.[9] FDIC identifies, monitors, and addresses risks to the deposit insurance fund when a bank or thrift institution fails.

The Bank Insurance Fund and the Savings Association Insurance Fund were established as FDIC responsibilities under the Financial Institutions Reform, Recovery, and Enforcement Act of 1989, which sought to reform, recapitalize, and consolidate the federal deposit insurance system.[10] The act also designated FDIC as the administrator of the Federal Savings & Loan Insurance Corporation Resolution Fund, which was created to complete the affairs of the former Federal Savings & Loan Insurance Corporation and liquidate the assets and liabilities transferred from the former Resolution Trust Corporation. The Bank Insurance Fund and the Savings Association Insurance Fund merged into the Deposit Insurance Fund on February 8, 2006, as a result of the passage of the Federal Deposit Insurance Reform Act of 2005.[11]

FDIC Relies on Computer Systems to Support Its Mission and Financial Reporting

FDIC relies extensively on computerized systems to support its mission, including financial operations, and to store the sensitive information that it collects. Its local and wide area networks interconnect these systems. To support its financial management functions, the corporation relies on many systems, including a corporatewide system that functions as a unified set of financial and payroll systems that are managed together and operated in an integrated fashion, a system to calculate and collect FDIC deposit insurance premiums and Financing Corporation bond principal and interest amounts from insured financial institutions;[12] a Web-based application that provides full functionality to support franchise marketing, asset marketing, and asset management; a system to request access to and receive permission for the computer applications and resources available to its employees, contractors, and other authorized personnel; and a primary receivership and subsidiary financial processing and reporting system. FDIC financial systems process and track financial transactions such as disbursements made to support operations. FDIC protects its computerized systems using a layered approach to security defense.

Under FISMA, the Chairman of FDIC is responsible for, among other things, (1) providing information security protections commensurate with the risk and magnitude of the harm resulting from unauthorized access, use, disclosure, disruption, modification, or destruction of the entity's information systems and information; (2) ensuring that senior agency officials provide information security for the information and information systems that support the operations and assets under their control; and (3) delegating to the corporation's Chief Information Officer the authority to ensure compliance with the requirements imposed on the agency under FISMA.

The Chief Information Officer is responsible for developing and maintaining a corporate wide information security program and for developing and maintaining information security policies, procedures, and control techniques that address all applicable requirements. The Chief Information Officer also serves as the authorizing official with the authority to approve the operation of the information systems at an acceptable level of risk to the corporation. The Chief Information Security Officer reports to the Chief Information Officer and serves as the Chief Information Officer's designated representative. The Chief Information Security Officer is responsible for the overall support of certification and accreditation activities.[13] According to FDIC policy, the Chief Information Security Officer is responsible for the development, coordination, and implementation of FDIC's security policy and the coordination of information security and privacy efforts across the corporation. The Chief Information Security Officer coordinates the process of building a corporate wide security strategy and vision to include the creation and maintenance of FDIC's information security policy, security risk assessment efforts, information technology risk assessments, disaster recovery, security monitoring, security awareness and training program, and security protection architecture.

Information Security Weaknesses Place Financial and Other Sensitive Information at Risk

FDIC did not sufficiently implement access and other controls intended to protect the confidentiality, integrity, and availability of its financial systems and information and other sensitive information. A key reason for these weaknesses is that FDIC did not always fully implement key information security program activities such as effectively developing and implementing security policies, and implementing an effective continuous monitoring program. These control deficiencies, which collectively constituted a significant deficiency for calendar year 2009, reduced FDIC's ability to ensure that authorized users had only the access needed to perform their assigned duties, and that its systems were sufficiently protected from unauthorized access. As a result, increased risk exists that financial information and other sensitive information could be disclosed or modified without authorization.

Access Controls to Information Resources Were Not Sufficient

A basic management objective for any organization is to protect the resources that support its critical operations and assets from unauthorized access. Organizations accomplish this by designing and implementing controls that are intended to prevent, limit, and detect unauthorized access to computer resources (e.g., data, programs, equipment, and facilities), thereby protecting them from unauthorized disclosure, modification, and loss. Specific access controls include authorization restrictions, system boundary protections, identification and authentication of users, cryptography, and audit and monitoring procedures. Without adequate access controls, unauthorized individuals, including intruders and former employees, can surreptitiously read and copy sensitive data and make undetected changes or deletions for malicious purposes or personal gain. In addition, authorized users can intentionally or unintentionally modify or delete data or execute changes that are outside of their authority.

User Access Was Not Sufficiently Restricted

Authorization is the process of granting or denying access rights and privileges to a protected resource, such as a network, system, application, function, or file. A key component of granting or denying access rights is the concept of "least privilege," which refers to granting users only the access rights and permissions that they need to perform their official duties. To restrict legitimate users' access to only those programs and files that they need in order to do their work, organizations establish user access rights: allowable actions that can be assigned to users or to groups of users. File and directory permissions are rules that are associated with a particular file or directory, regulating which users can access it—and the extent of their access rights. To avoid unintentionally giving a user unnecessary access to sensitive files and directories, an organization should give careful consideration to its assignment of rights and permissions. In addition, National Institute of Standards and Technology (NIST) guidance states that an organization should enforce approved authorizations for logical access to its information systems with access mechanisms such as access control lists within the network system.[14] Furthermore, NIST guidance states that access should be allowed only for authorized users and only for the tasks necessary to accomplish their work in accordance with the organization's missions and business functions.

FDIC did not always sufficiently restrict system access and privileges to only those users who needed access to perform their assigned duties. For example, FDIC did not always

- configure access control lists on all its network devices to limit or restrict network traffic,
- configure access control lists on servers dedicated to network management to restrict access to only those users who require it,
- ensure access to sensitive files of critical network devices was adequately controlled,
- control access to a database supporting an accounting application used to process receivership asset financial activity, and
- limit user access rights to only those roles necessary to perform their duties.

As a result, increased risk exists that a user could gain inappropriate access to computer resources, circumvent security controls, and deliberately or inadvertently read, modify, or delete financial information and other sensitive information.

System Boundary Protections Were Not Adequately Enforced

Boundary protection controls logical connectivity into and out of networks and controls connectivity to and from network-connected devices. Unnecessary connectivity to an organization's network increases not only the number of access paths that must be managed and the complexity of the task, but also the risks of unauthorized access. National Security Agency (NSA) guidelines state that, to reduce the probability of a successful network penetration, the data and telephony networks must be logically separated.

FDIC did not control access to its data network by separating or partitioning the data network from the voice network. Physical convergence of voice and data networks is an advantage of Internet Protocol (IP) telephony systems; however, placing both systems on the same network means both are now susceptible to the same attacks and the same attackers. As a result, increased risk exists that unauthorized or malicious users could gain access to the data network and inadvertently read, modify, or delete financial information and other sensitive information.

Identification and Authentication User Controls Were Not Consistently Enforced

A computer system must be able to identify and authenticate the identity of a user so that activities on the system can be linked to that specific individual and to protect its systems from inadvertent or malicious access. When an organization assigns unique user accounts to specific users, the system is able to distinguish one user from another—a process called identification. This allows the system to distinguish and track one user from another. The system must also establish the validity of the user's claimed identity by requesting some kind of information, such as a password, that is known only by the user—a process known as authentication. The NSA security guidelines state that standard or default community strings should not be used.[15] In addition, NIST guidance states that an organization should manage information system authenticators by changing the default content of authenticators (e.g., passwords) when installing an information system. Furthermore, FDIC policy states that passwords should be changed after 90 days.

FDIC did not consistently enforce identification and authentication controls for its users and systems. Specifically,

- FDIC had not securely configured the Simple Network Management Protocol (SNMP) community strings.[16] Also, FDIC had not recently changed the SNMP read-write community string for administering routers and switches.
- An FDIC network software package was operating with a default vendor- supplied identification (ID) and password.
- Several service and administrator user accounts on UNIX servers were not required to change their passwords in accordance with FDIC policy or were set to never expire.

As a result of these weaknesses, increased risk exists that a user would not be uniquely identified before accessing the FDIC network, leaving FDIC without a reliable trail to follow to hold the user accountable in the event of a security incident.

Sensitive Information Was Not Always Encrypted

Cryptography underlies many of the mechanisms used to enforce the confidentiality and integrity of sensitive information. A basic element of cryptography is encryption.[17] Encryption can be used to provide basic data confidentiality and integrity by transforming plain text into cipher text using a special value known as a key and a mathematical process known as an algorithm.[18] If encryption is not used, user ID and password combinations will be susceptible to electronic eavesdropping by devices on the network when they are transmitted. NSA and NIST recommend encrypting network services, and NIST guidance states that organizations should configure an information system to provide only the essential capabilities needed or restrict the use of protocols that can allow the unauthorized transfer of information.[19] NIST guidance also states that the use of encryption by organizations can reduce the probability of unauthorized disclosure of information.

FDIC did not always ensure that sensitive information transmitted over its network was adequately encrypted. Specifically, FDIC did not disable an unencrypted protocol in use in the production and nonproduction logical partitions on the mainframe. In addition, FDIC did not restrict the use of unencrypted protocols on network servers. As a result, increased risk exists that an individual could capture information such as user IDs and passwords and use them to gain unauthorized access to data and system resources.

Audit and Monitoring of Security-Relevant Events Were Inadequate

To establish individual accountability, monitor compliance with security policies, and investigate security violations, it is crucial to determine what, when, and by whom specific actions have been taken on a system. Organizations accomplish this by implementing system or security software that provides an audit trail for determining the source of a transaction or attempted transaction and monitoring user activity. To be effective, organizations should (1) configure the software to collect and maintain a sufficient audit trail for security-relevant events; (2) generate reports that selectively identify unauthorized, unusual, and sensitive access activity; and (3) regularly monitor and take action on these reports. NIST guidance states that an organization should review and analyze information system audit records for indications of inappropriate or unusual activity, and should report the findings to designated organization officials. NIST guidance also states that organizations should track and monitor access by individuals who use elevated access privileges and that an organization should review and analyze information system audit records for indications of inappropriate or unusual activity, and should report the findings to designated organization officials.

FDIC did not always sufficiently review audit and monitoring of security-relevant events. For example,

- FDIC's monitoring processes did not detect the existence of default installation user accounts on three UNIX servers.
- FDIC did not effectively monitor certain dataset access activity on the mainframe.
- FDIC mainframe logging controls were inappropriately configured, allowing the creation of large quantities of logged data for routine activities.

As a result of these deficiencies, increased risk exists that unauthorized activity or a policy violation would not be detected on FDIC systems and networks.

Weaknesses in Other Information System Controls Increased Risk

In addition to access controls, other important controls should be in place to ensure the confidentiality, integrity, and availability of an organization's information. These controls include policies, procedures, and techniques for securely segregating incompatible duties, configuring information systems, and updating continuity documents. However, FDIC weaknesses in these areas have increased the risk of unauthorized use, disclosure, modification, or loss of information and information systems.

Incompatible Duties and Functions Were Not Adequately Segregated

In addition to having access controls, an organization should have policies, procedures, and controls in place to appropriately segregate computer- related duties. Segregation of duties refers to the policies, procedures, and organizational structure that help ensure that one individual cannot independently control all key aspects of a process or computer-related operation and thereby gain unauthorized access to assets or records. Often segregation of incompatible duties is achieved by dividing responsibilities among two or more organizational groups, which diminishes the likelihood that errors and wrongful acts will go undetected because the activities of one individual or group will serve as a check on the activities of the other. Inadequate segregation of duties increases the risk that erroneous or fraudulent transactions could be processed, improper program changes implemented, and computer resources damaged or destroyed. FDIC policy on UNIX security states that development and production data shall be separated and access controlled such that application developers have access to only development areas (nonproduction) and application users have access to only production areas (nondevelopment).

FDIC did not always adequately segregate incompatible computer-related duties and functions. For example, FDIC developers are allowed access to both development and production UNIX servers. Allowing developers such access reduced FDIC's ability to achieve segregation of duties and increased the risk of unauthorized and unnecessary access to sensitive production data by individuals who do not require it to perform their job duties. As a result, increased risk exists that users could perform unauthorized system activities without detection.

Although Elements of Configuration Management Controls Existed, They Were Not Always Fully Implemented

Configuration management is another important control that involves the identification and management of security features for all hardware and software components of an information system at a given point and systematically controls changes to that configuration during the system's life cycle. An effective configuration management process includes procedures for identifying, documenting, and assigning unique identifiers (for example, serial number and name) to a system's hardware and software parts and subparts, generally referred to as configuration items, and evaluating and deciding whether to approve changes to a system's baseline configuration. In addition, vendor specification states that engineers will not develop, repair, maintain, or test the product software after a system reaches its end-of-life date. NIST guidance also states that an organization should promptly install security-relevant software updates, such as patches.

FDIC has implemented some elements of a configuration management process. Specifically, it has documented policy and procedures for assigning unique identifiers and naming configuration items so that they can be distinguished from one another and for requesting changes to configuration items. FDIC has also developed a change request process and a baseline for its systems.

However, FDIC did not always implement key configuration management controls over its information system components. For example, FDIC had critical end-of-life systems that were not supported by their manufacturers, which indicates that patches or updates for emerging threats were no longer available. In addition, patch levels for third-party software running on two UNIX servers at FDIC were not current and an obsolete version of third-party software was running on a Windows server. As a result, increased risk exists that these FDIC systems will be exposed to unauthorized access or manipulation through the exploitation of known vulnerabilities that have not been patched or emergent vulnerabilities for which no known remedy exists.

Contingency Planning Documentation Was Incomplete or Had Incorrect Information

Contingency planning, which includes developing contingency, business continuity, and disaster recovery plans, should be performed to ensure that when unexpected events occur, essential operations can continue without interruption or can be promptly resumed, and that sensitive data are protected. NIST guidance states that organizations should develop and implement a contingency plan that describes activities associated with backing up and restoring the system after a disruption or failure. The plan should be updated and include information such as contact, resources, and description of files in order to restore the application in the event of a disaster. In addition, the plans should be tested to determine the plans' effectiveness and the organization's readiness to execute the plans. Officials should review the plan results and initiate corrective actions.

FDIC has developed contingency plans, business continuity plans, and disaster recovery plans and has also conducted testing on these plans. However, one contingency plan at the Virginia Square office contained incomplete information. Specifically, the contingency plan did not include information such as resources (servers, applications, network components, supplies, telecommunications, and databases) and technical information about databases, libraries, and guidance for restoring devices to operational order.

In addition, the business continuity plans at the Dallas office did not have correct information. Specifically, one of the plans was dated September 2005, and was created for a previous location. In addition, another plan had missing information. It listed a contact person for the headquarters security department who no longer worked there. Also, there was no contact information for the emergency management team.

At the time of our review, FDIC officials stated that they will update the contingency and continuity plans with the required information. Until FDIC maintains current contingency and continuity plans, increased risk exists that it will not be able to effectively recover and continue operations when an emergency occurs and its operations will be disrupted.

FDIC Had Not Fully Implemented Its Information Security Program

A key reason for the information security weaknesses is that although FDIC has made important progress in implementing its security program, it did not always fully complete key information security program activities. FDIC has provided employees with security awareness and security-specific training, and has implemented a system to track remedial action plans to ensure that deficiencies are mitigated in an effective and timely manner. However, FDIC did not always fully implement key information security program activities such as effectively developing and implementing security policies, and implementing an effective continuous monitoring program. Until all key elements of its information security program have been fully and consistently implemented, FDIC will not have sufficient assurance that its financial information and assets are adequately safeguarded from inadvertent or deliberate misuse, fraudulent use, improper disclosure, or destruction.

Security Policies and Procedures Were Not Always Developed, Documented, or Implemented

A key task in developing an effective information security program is to establish and implement risk-based policies, procedures, and technical standards that govern security over an agency's computing environment. If properly implemented, policies and procedures help reduce the risk that could come from unauthorized access or disruption of services. Because security policies and procedures are the primary mechanisms through which management communicates its views and requirements, it is important that these policies and procedures be established and documented. FISMA requires agencies to develop and implement policies and procedures to support an effective information security program. NIST has also issued security standards and related guidance to help entities implement security controls, including appropriate information security policies and procedures.

While FDIC has generated agencywide information security policy relating to access control and risk management, certain policies and procedures had not always been developed, documented, and implemented. For example,

- FDIC did not develop or document policies and procedures to prevent users from having inappropriate or incompatible access to multiple applications. For example, FDIC did not have policies and procedures to identify and govern the assignment of access privileges to combinations of systems that create logical access to data that is otherwise prevented by applications. As a result, a combination of access privileges assigned to individuals allowed for the circumvention of an accounting application's access controls. Additionally, FDIC did not develop or document technical controls in place to identify or prevent the assignment of such combinations of access privileges that expose the data associated with certain applications from access outside of the access controls implemented within the functions of those applications. As a result, individuals could inappropriately obtain access to data in certain applications.

- FDIC did not implement the policy requiring service and administrator user accounts on UNIX servers to change their passwords.
- FDIC did not implement the policy to have many of the mainframe IDs set with an "expire date."

Until these policies and procedures are fully developed, documented, and implemented, FDIC has reduced assurance that computing resources are consistently and effectively protected from inadvertent or deliberate misuse, including fraud or destruction.

Continuous Monitoring Efforts Were Not Always Sufficient

NIST states that a continuous monitoring program allows an organization to maintain the security authorization of an information system over time in a highly dynamic environment of operation with changing threats, vulnerabilities, technologies, and business processes. Continuous monitoring of security controls using automated support tools facilitates near real-time risk management and promotes organizational situational awareness with regard to the state of the security of the information system. The implementation of a continuous monitoring program can result in ongoing updates to the security plan, the security assessment report, and the plan of action and milestones, the three principal documents in the security authorization package. A rigorous and well- executed continuous monitoring program significantly reduces the level of effort required for the reauthorization of the information system. FDIC policy states that in addition to performing system test and evaluation in support of ongoing certification and accreditation efforts, FDIC is to routinely test major applications and their components as part of the continuous monitoring program. The program is designed to identify the most commonly exploited application-level vulnerabilities that exist within the enterprise infrastructure.

FDIC's continuous monitoring efforts were not always sufficient. FDIC did not sufficiently (1) monitor users' inappropriate and excessive access privileges to a business application that supports resolution and receivership activities, (2) have the ability to reliably detect changes to powerful mainframe programs, and (3) test and verify that all system interfaces were properly configured for the new systems before putting them into production. FDIC has a continuous monitoring process; however, several of the vulnerabilities we identified with respect to FDIC's security over its information systems were not identified through FDIC's routine monitoring of access privileges, audit logs, and adherence to established policies and procedures. We identified numerous access control vulnerabilities that were not identified by the continuous monitoring program. These vulnerabilities resulted in significant reductions in FDIC's capability to maintain effective controls and to protect the confidentiality, integrity, and availability of its information systems and information. As a result, FDIC had limited assurance that computing resources were being consistently and effectively protected from inadvertent or deliberate misuse, including fraud or destruction.

FDIC Has Mitigated Previously Reported Weaknesses

Despite the newly identified weaknesses, FDIC has made progress in mitigating previously reported information security weaknesses. The corporation has mitigated all 10 of the information security weaknesses reported in our calendar year 2007 audit.[20] To its credit, the corporation has made improvements to the configuration management controls and aspects of its security management. For example, it maintained a full and complete requirements baseline for two systems and included key information in a remedial action plan. In addition, FDIC has corrected the FDIC Inspector General findings from the 2009 report.[21]

Conclusions

FDIC had many new control weaknesses putting its systems at a higher level of vulnerability to internal threats. These weaknesses impair the corporation's ability to ensure the confidentiality, integrity, and availability of financial and sensitive information. The weaknesses also represent a significant deficiency in internal controls over the information systems and data used for financial reporting. Despite the newly identified weaknesses, FDIC has made progress in mitigating previously reported information security weaknesses.

Until FDIC (1) mitigates known information security weaknesses in access controls and other information system controls and (2) fully implements a comprehensive agencywide information security program that includes developing, documenting, and implementing security policies and implementing an effective continuous monitoring program, its financial and other sensitive information will remain at increased risk of unauthorized disclosure, modification, or destruction, and its management decisions may be based on unreliable or inaccurate information.

RECOMMENDATIONS FOR EXECUTIVE ACTION

We recommend that the Chairman direct the Chief Information Officer to take the following two actions to enhance the corporation's information security program:

- develop and document policies and procedures for assigning access to systems and databases where application controls could be compromised, and
- complete the implementation of an effective continuous monitoring program to detect vulnerabilities.

We are also making 31 new recommendations to address 28 new findings in a separate report with limited distribution. These recommendations consist of actions to implement and correct specific information security weaknesses related to access controls, segregation of duties, configuration management, and contingency planning identified during this audit.

AGENCY COMMENTS

In providing written comments on a draft of this report, the Deputy to the Chairman and Chief Financial Officer of FDIC generally agreed with our recommendations. In addition, the Deputy discussed the actions that FDIC has taken or plans to take to implement the recommendations, such as restricting access to systems and databases and enhancing its continuous monitoring program as part of an ongoing multiyear effort.

Gregory C. Wilshusen
Director, Information Security Issues

Dr. Nabajyoti Barkakati
Director, Chief Technologist

APPENDIX I. OBJECTIVES, SCOPE, AND METHODOLOGY

The objectives of our review were to assess (1) the effectiveness of the Federal Deposit Insurance Corporation's (FDIC) controls in protecting the confidentiality, integrity, and availability of its financial systems and information, and (2) the progress FDIC has made in mitigating previously reported information security weaknesses. These objectives were integral to supporting our opinion on FDIC's internal controls provided in conjunction with our integrated audit of the financial statements of the two funds administrated by FDIC, by assessing the controls over systems that support financial management and the generation of financial statements for two FDIC funds.

To determine whether controls over key financial systems were effective, we tested the effectiveness of information security and information technology-based internal controls. We concentrated our evaluation primarily on the controls for financial applications and enterprise database applications associated with a corporatewide system that functions as a unified set of financial and payroll systems that are managed together and operated in an integrated fashion; a system to calculate and collect FDIC deposit insurance premiums and Financing Corporation bond principal and interest amounts from insured financial institutions;[22] a Web-based application that provides full functionality to support franchise marketing, asset marketing, and asset management; a system to request access to and receive permission for the computer applications and resources available to its employees, contractors, and other authorized personnel; a primary receivership and subsidiary financial processing and reporting system; and the general support systems. Our selection of the systems was based on discussions with our stakeholders.

Our evaluation was based on our *Federal Information System Controls Audit Manual*, which contains guidance for reviewing information system controls that affect the confidentiality, integrity, and availability of computerized information.

Using National Institute of Standards and Technology (NIST) standards and guidance and FDIC's policies, procedures, practices, and standards, we evaluated controls by

- observing methods for providing secure data transmissions across the network to determine whether sensitive data were being encrypted;
- testing and observing physical access controls to determine if computer facilities and resources were being protected from espionage, sabotage, damage, and theft;
- evaluating the control configurations of selected servers and database management systems;
- inspecting key servers and workstations to determine whether critical patches had been installed or were up-to-date; and
- examining access responsibilities to determine whether incompatible functions were segregated among different individuals.

Using the requirements of the Federal Information Security Management Act (FISMA), which establishes key elements for an effective agencywide information security program, we evaluated FDIC's implementation of its security program by

- reviewing FDIC's risk assessment process and risk assessments for key FDIC systems that support the preparation of financial statements to determine whether risks and threats were documented consistent with federal guidance;
- analyzing FDIC's policies, procedures, practices, and standards to determine their effectiveness in providing guidance to personnel responsible for securing information and information systems;
- analyzing security plans to determine if management, operational, and technical controls were in place or planned and that security plans were updated;
- examining training records for personnel with significant security responsibilities to determine if they had received training commensurate with those responsibilities;
- analyzing configuration management plans and procedures to determine if configurations were being managed appropriately;
- analyzing security testing and evaluation results for four key FDIC systems to determine whether management, operational, and technical controls were tested at least annually and based on risk;

- examining remedial action plans to determine whether they addressed vulnerabilities identified in FDIC's security testing and evaluations; and
- examining contingency plans for four key FDIC systems to determine whether those plans had been tested or updated.

We also discussed with key security representatives and management officials whether information security controls were in place, adequately designed, and operating effectively.

To determine the status of FDIC's actions to correct or mitigate previously reported information security weaknesses, we identified and reviewed its information security policies, procedures, and guidance. We reviewed prior GAO reports to identify previously reported weaknesses and examined FDIC's corrective action plans to determine which weaknesses FDIC had reported were corrected. For those instances where FDIC reported it had completed corrective actions, we assessed the effectiveness of those actions.

We conducted this performance audit from December 2009 to November 2010 in accordance with generally accepted government auditing standards. We performed our data collection, analysis, and assessment procedures in support of the financial audit during the December 2009 to June 2010 time frame. We performed supplemental audit procedures to prepare this report from June 2010 to November 2010. Those standards require that we plan and perform the audit to obtain sufficient, appropriate evidence to provide a reasonable basis for our findings and conclusions based on our audit objectives. We believe that the evidence obtained provides a reasonable basis for our findings and conclusions based on our audit objectives.

End Notes

[1] Information system general controls affect the overall effectiveness and security of computer operations and are not unique to specific computer applications. These controls include security management, configuration management, operating procedures, software security features, and physical protections designed to ensure that access to data is appropriately restricted, that only authorized changes to computer programs are made, that incompatible computer-related duties are segregated, and that backup and recovery plans are adequate to ensure the continuity of operations.

[2] GAO, *Financial Audit: Federal Deposit Insurance Corporation Funds' 2009 and 2008 Financial Statements*, GAO-10-705 (Washington, D.C.: June 25, 2010).

[3] A significant deficiency is a control deficiency, or combination of deficiencies, in internal control that is less severe than a material weakness, yet important enough to merit attention by those charged with governance. A deficiency in internal control exists when the design

or operation of a control does not allow management or employees, in the normal course of performing their assigned functions, to prevent, or detect and correct, misstatements on a timely basis.

[4] We performed data collection, analysis, and assessment procedures in support of the financial audit during the December 2009 to June 2010 time frame. We performed supplemental audit procedures to prepare this report from June 2010 to November 2010.

[5] The Department of Homeland Security's (DHS) federal information security incident center is hosted by US-CERT. When incidents occur, agencies are to notify the center.

[6] GAO, *High-Risk Series: Information Management and Technology*, GAO/HR-97-9 (Washington, D.C.: February 1997), and *High-Risk Series: An Update*, GAO-09-271 (Washington, D.C.: January 2009).

[7] FISMA was enacted as title III, E-Government Act of 2002, Pub. L. No. 107-347, 116 Stat. 2899, 2946 (Dec. 17, 2002).

[8] Federal Deposit Insurance Corporation Act, June 16, 1933, Ch. 89, § 8.

[9] FDIC is an independent agency of the federal government and receives no direct congressional appropriations; it is funded by premiums that banks and thrift institutions pay for deposit insurance coverage and from earnings on investments in U.S. Treasury securities.

[10] Pub. L. No. 101-73, § 211, 103 Stat. 183, 218-22 (Aug. 9, 1989).

[11] Pub. L. No. 109-171, Title II, Subtitle B, § 2102 (Feb. 8, 2006).

[12] The Financing Corporation, established by the Competitive Equality Banking Act of 1987, is a mixed-ownership government corporation with its primary purpose being to function as a financing vehicle for the Federal Savings & Loan Insurance Corporation. Effective December 12, 1991, as provided by the Resolution Trust Corporation Refinancing, Restructuring and Improvement Act of 1991, the Financing Corporation's ability to issue new debt was terminated. Outstanding Financing Corporation bonds, which are 30-year non-callable bonds with a principal amount of approximately $8.1 billion, mature in 2017 through 2019.

[13] The Office of Management and Budget (OMB) requires that a management official formally authorize (or accredit) an information system to process information and accept the risk associated with its operation based on a formal evaluation (or certification) of the system's security controls. For annual reporting, OMB requires agencies to report the number of systems, including impact levels, authorized for processing after completing certification and accreditation.

[14] Logical access requires users to authenticate themselves (through the use of secret passwords or other identifiers) and limit the files and other resources that authenticated users can access and the actions that they can execute.

[15] The community string (also known as the community name) provides a weak authentication mechanism to the Simple Network Management Protocol (SNMP). Agents can be configured to allow read-only, read-write, or no access to their parameters based on the community string in a request. Community strings are passed in clear text in SNMP messages, so they can be easily sniffed and are therefore insufficient for authenticating legitimate manager requests.

[16] SNMP enables network and system administrators to remotely monitor and configure devices on the network (devices such as switches and routers).

[17] Encryption is a subset of cryptography, which is used to secure transactions by providing ways to ensure data confidentiality (assurance that the information will be protected from unauthorized access), data integrity (assurance that data have not been accidentally or deliberately altered), authentication of the message's originator, electronic certification of data, and nonrepudiation (proof of the integrity and origin of data that can be verified by a third party).

[18] A cryptographic algorithm and key are used to apply cryptographic protection to data (e.g., encrypt the data or generate a digital signature) and to remove or check the protection (e.g., decrypt the encrypted data or verify the digital signature).

[19] A protocol is a set of rules or procedures for transmitting data between electronic devices, such as computers. In order for computers to exchange information, there must be a preexisting agreement as to how the information will be structured and how each side will send and receive it.

[20] GAO, *Information Security: FDIC Sustains Program but Needs to Improve Configuration Management of Key Financial Systems*, GAO-08-564 (Washington, D.C.: May 30, 2008).

[21] FDIC Office of Inspector General, *Information Technology Controls in Support of the FDIC Fund's 2008 and 2007 Financial Statement Audit*, AUD-09-020 (Washington, D.C.: Aug. 17, 2009).

[22] The Financing Corporation, established by the Competitive Equality Banking Act of 1987, is a mixed-ownership government corporation with its primary purpose being to function as a financing vehicle for the Federal Savings & Loan Insurance Corporation. Effective December 12, 1991, as provided by the Resolution Trust Corporation Refinancing, Restructuring and Improvement Act of 1991, the Financing Corporation's ability to issue new debt was terminated. Outstanding Financing Corporation bonds, which are 30-year non-callable bonds with a principal amount of approximately $8.1 billion, mature in 2017 through 2019.

In: Federal Information Security
Editors: D. G. Miller and I. I. Wilson

ISBN 978-1-61942-771-6

Chapter 3

INFORMATION SECURITY: NATIONAL NUCLEAR SECURITY ADMINISTRATION NEEDS TO IMPROVE CONTINGENCY PLANNING FOR ITS CLASSIFIED SUPERCOMPUTING OPERATIONS

United States Government Accountability Office

WHY GAO DID THIS STUDY

In the absence of underground nuclear weapons testing, the National Nuclear Security Administration (NNSA) relies on its supercomputing operations at its three weapons laboratories to simulate the effects of changes to current weapons systems, calculate the confidence of future untested systems, and ensure military requirements are met.

GAO was requested to assess the extent to which (1) NNSA has implemented contingency and disaster recovery planning and testing for its classified supercomputing systems, (2) the laboratories are able to share supercomputing capacity for recovery operations, and (3) NNSA tracks the costs for contingency and disaster recovery planning for supercomputing assets. To do this work, GAO examined contingency and disaster recovery planning policies and activities, and analyzed classified supercomputing capabilities at the weapons laboratories, and NNSA budgetary data.

What GAO Recommends

GAO recommends, among other things, that NNSA clearly define roles and responsibilities for its component organizations in providing oversight for contingency and disaster recovery planning for the classified supercomputing environment. NNSA agreed with most of GAO's recommendations, but did not concur with recommendations relating to capacity planning and cost tracking.

What GAO Found

All three NNSA weapons laboratories—Los Alamos, Sandia, and Lawrence Livermore—have implemented some components of a contingency planning and disaster recovery program. NNSA, however, has not provided effective oversight to ensure that the laboratories have comprehensive and effective contingency and disaster recovery planning and testing. Further, due to lack of planning and analysis by NNSA and the laboratories, the impact of a system outage is unclear. Only one of the three laboratories—Los Alamos—had conducted a business impact analysis to assess the criticality of resources and acceptable outage time frames; yet, NNSA and all three laboratories consider the consequence associated with the loss of system availability to be low impact and do not consider the classified supercomputers to be mission critical. Nonetheless, NNSA classified supercomputing capabilities serve as a computational surrogate to nuclear weapons testing and are used to address other areas of national security. Despite the absence of business impact analyses, all laboratories had key components of a contingency planning program in place. However, shortcomings existed. For example, all laboratories had backup processes in place and had developed contingency plans, but the plans were not comprehensive. Specifically, one plan did not address the supercomputing operations, and none of the plans had been tested at the time of GAO's review. In addition, the laboratories addressed disaster recovery to a limited extent, but not specifically for the supercomputers. These shortcomings existed, at least in part, because NNSA's component organizations, including the Office of the Chief Information Officer, were unclear about their roles and responsibilities for providing oversight in the laboratories' implementation of contingency and disaster recovery planning. Until the agency fully implements a contingency and disaster recovery planning program for its weapons laboratories, it has limited assurance that vital information can be recovered and made available to meet national security priorities and requirements.

Although the laboratories have the technological capability to share supercomputing capacity across all three weapons laboratories, barriers exist that could impede recovery operations. For example, the laboratories do not know the minimum supercomputing capacity needed to meet program requirements, such as simulating the effects of changes to weapons systems, should a disruption occur. In addition, the laboratories have not tested the technological capability to share the capacity on an on-demand basis for recovery operations. Without having an understanding of capacity needs and subsequent testing, the laboratories have little assurance that they could effectively share capacity if needed.

Although NNSA obligated approximately $1.7 billion to help implement its classified supercomputing program from fiscal years 2007 through 2009, the agency has not tracked costs for contingency and disaster recovery planning and is uncertain of actual funds that were spent toward these efforts.

ABBREVIATIONS

ASC	Advanced Simulation and Computing
BIA	business impact analysis
CNSS	Committee on National Security Systems
DISCOM	Distance Computing
DOD	Department of Defense
DOE	Department of Energy
FISMA	Federal Information Security Management Act of 2002
FLOPS	floating-point operations per second
Livermore	Lawrence Livermore National Laboratory
Los Alamos	Los Alamos National Laboratory
NIST	National Institute of Standards and Technology
NNSA	National Nuclear Security Administration
Sandia	Sandia National Laboratories

December 9, 2010

The Honorable Henry Waxman
Chairman

Committee on Energy and Commerce
House of Representatives

The Honorable Edward J. Markey
Chairman
Subcommittee on Energy and the Environment
Committee on Energy and Commerce
House of Representatives

The Honorable Bart Stupak
Chairman
Subcommittee on Oversight and Investigations
Committee on Energy and Commerce
House of Representatives

The National Nuclear Security Administration[1] (NNSA) provides classified supercomputing capabilities for assessing the performance of nuclear weapons. In the absence of nuclear weapons testing—which ceased in 1992—the simulation capabilities of NNSA's supercomputers are a necessary means to determine the effects of changes to current weapons systems and to determine a level of confidence in the performance of future untested systems.[2] These simulation capabilities also contribute to the enhancement of NNSA's ability to predict the performance of weapons systems to ensure the systems meet all military requirements established by the Department of Defense (DOD).

NNSA's three nuclear weapons laboratories—Los Alamos National Laboratory (Los Alamos) in New Mexico, Lawrence Livermore National Laboratory (Livermore) in California, and the Sandia National Laboratories (Sandia) with locations in New Mexico and California—use these supercomputing simulation capabilities to obtain a comprehensive understanding of the entire nuclear weapons life cycle, from design to safe processes for dismantlement. These classified supercomputing capabilities are a considerable investment and serve as a cornerstone for NNSA's Stockpile Stewardship Program.[3] In addition, classified supercomputing capabilities are essential for informing critical decisions related to the nuclear stockpile, including all stockpile modernization and warhead studies. NNSA classified supercomputing capabilities are also used to address other areas of national security, including intelligence analyses, nuclear forensics, and emergency response. Because of the importance of these classified supercomputing capabilities to issues central to national security, contingency and disaster recovery planning[4] are key to ensuring that, when unexpected events occur, NNSA can recover and reconstitute its classified supercomputing systems, data, and operations.

Our objectives were to assess the extent to which (1) NNSA has implemented contingency and disaster recovery planning and testing for its classified supercomputing assets, (2) the three laboratories are able to share classified supercomputing capacity for recovery operations, should service disruptions occur, and (3) NNSA tracks the costs for ensuring contingency and disaster recovery planning for its classified supercomputing assets. To accomplish these objectives, we examined contingency and disaster recovery planning controls for the systems within the classified supercomputing environment that are a necessary means for NNSA's achievement of its nuclear weapons mission. In addition, we performed technical assessments of classified supercomputing capabilities at each weapons laboratory, including each laboratory's ability to share supercomputing capacity. Further, we obtained information from NNSA and laboratory officials to determine how expenditures were tracked for contingency and disaster recovery planning of the classified supercomputing systems at each of the laboratories, as well as projected future cost estimates for ensuring the recovery of these assets.

We conducted this performance audit from December 2009 through December 2010 in accordance with generally accepted government auditing standards. Those standards require that we plan and perform the audit to obtain sufficient, appropriate evidence to provide a reasonable basis for our findings and conclusions based on our audit objectives. We believe that the evidence obtained provides a reasonable basis for our findings and conclusions based on our audit objectives. A more detailed description of our objectives, scope, and methodology is contained in appendix I.

BACKGROUND

NNSA relies on its Stockpile Stewardship Program to ensure the safety, security, and effectiveness of the nuclear weapons stockpile. The Stockpile Stewardship Program is comprised of various elements, including, but not limited to: (1) the Advanced Simulation and Computing (ASC) Campaign, which provides the computational science and simulation tools to understand the behaviors and effects of nuclear weapons; (2) Directed Stockpile Work, which provides evidence of the health of the nuclear weapons stockpile and involves day-to-day maintenance of these weapons, including life extension efforts; (3) the Science Campaign, which provides tools and capabilities geared toward advancing the general understanding of all nuclear weapons systems; and (4) the Engineering Campaign, which provides a sustained basis for stockpile certification and assessments throughout the life cycle of each weapon. The coordination among the Stockpile Stewardship elements is instrumental to increasing NNSA's confidence in the performance of nuclear weapons.

To help accomplish its Stockpile Stewardship mission, NNSA relies on the three weapons laboratories—Los Alamos, Livermore, and Sandia. Los Alamos and Livermore are the two design laboratories that are responsible for designing the nuclear weapons' explosive package and conducting research to better understand nuclear weapons phenomena. Sandia is an engineering laboratory and has principal responsibility for the research, design, and development of nonnuclear warhead components; integration of these components with Los Alamos and Livermore; and overall warhead systems integration with DOD. In accordance with NNSA, management and operations contractors, who are responsible for day-to-day operations of the laboratories, are required to adhere to agency policies.[5]

At the time of our review, NNSA's classified supercomputing resources consisted of 12 classified supercomputing systems. Figure 1 shows the hardware configuration of a supercomputing system.

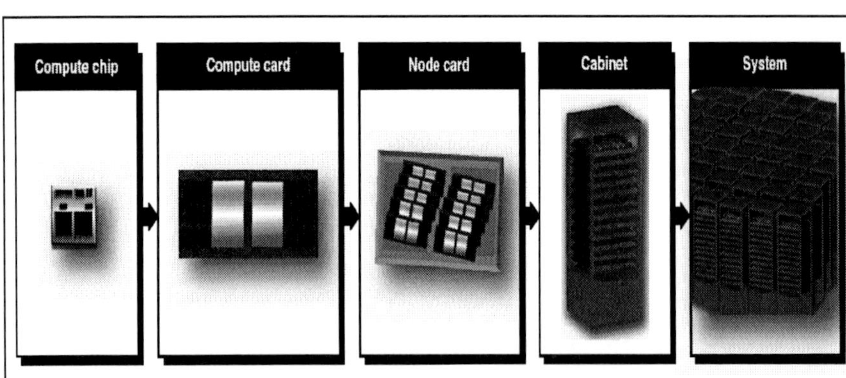

Source: GAO, data provided by Los Alamos, Livermore, and Sandia.

Figure 1. Common Hardware Components of a Supercomputing System.

NNSA classified supercomputing systems employ a large number of interdependent processors, which are the core unit of a computer that gathers instructions and data. These processors are mounted onto a compute chip, which is the portion of the system that carries out the instructions of a computer program. These compute chips are inserted onto a compute card, which also holds memory for the compute chips to use. A number of compute cards are attached to a node card, which have one or more processors with a common memory and are connected by high-speed interconnection networks. Each node card is inserted into a single cabinet, and that configuration is repeated many times to build a single supercomputing system. Each supercomputing system has a peak performance, which is the maximum rate of floating-point operations per second (FLOPS) that the system can sustain.[6] Currently, almost all NNSA classified supercomputer systems operate at the teraFLOP level, which represents a trillion FLOPS.

According to NNSA, the laboratories have three types of classified supercomputing systems:

Capacity: Small systems that execute parallel problems with more modest computational requirements. These systems serve as the workhorse for the ASC program and are responsible for processing the day-to-day supercomputing workload.

Capability: This type of supercomputer is used to solve the largest and most demanding problems that other computing systems cannot manage.

Advanced architecture: Research and development systems that assist the ASC program in preparing to rapidly deploy and exploit the next generation of supercomputing technology. These systems have a targeted workload and serve as the foundation for the next generation of NNSA supercomputers.

Table 1 shows the classified supercomputing systems currently in use at the three weapons laboratories.

NNSA's classified supercomputing capabilities consist of supporting resources, including (1) parallel files systems, which store transitory data; (2) network file systems, which store user and project data for a calculation; (3) archival storage systems, which serve as storage for data; and (4) visualization systems, which enable users to better comprehend the results of their computations.

Table 1. Inventory of NNSA-Deployed Classified Supercomputing Systems (as of October 2010)

Site	System name	System type	Delivery date	Total processors	Peak performance (TeraFLOPS)
Los Alamos	Roadrunner Base	Capacity	10/2006	18,432	76.0
	Roadrunner Phase-3	Advanced architecture	9/2008	24,480	1,280.0
	Hurricane	Capacity	9/2008	5,760	51.2
Livermore	BlueGene/L	Advanced architecture	11/2004	131,072	367.0
	Purple[a]	Capability	6/2005	12,288	93.4
	Rhea	Capacity	9/2006	4,608	22.1
	Minos	Capacity	6/2007	6,912	33.2
	Juno	Capacity	5/2008	18,432	162.2
	Dawn	Advanced architecture	1/2009	147,456	501.4
Sandia-NM	Red Storm	Advanced architecture	3/2005	31,680	284.0
	Unity	Capacity	3/2009	4,352	38.0
Sandia-CA	Whitney	Capacity	3/2009	4,352	38.0

Source: GAO summary of data from Los Alamos National Laboratory, Lawrence Livermore National Laboratory, and Sandia National Laboratories.

[a] Although Purple was the capability system in use at the time of our site visits, Livermore retired the system in November 2010.

NNSA's classified supercomputing systems are connected via its Enterprise Secure Network and the Distance Computing (DISCOM) network, which function as supporting resources for the classified supercomputing environment. The Enterprise Secure Network provides classified communications across the nuclear weapons complex, including security services and other activities that ensure the flow of NNSA's data sharing and business missions. DISCOM provides secure, high-speed remote access for intra- and inter-site file transfers and enables users, across the three weapons laboratories, to operate on remote computing resources as if they were local. DISCOM and the Enterprise Secure Network serve as the backup networks to each other. Figure 2 shows the composition of NNSA's classified supercomputing network infrastructure.

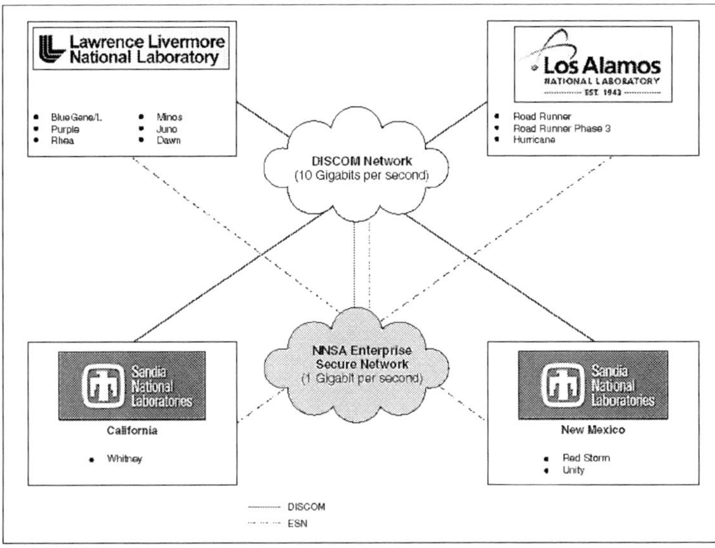

Source: GAO, data provided by Los Alamos, ivermore, and Sandia.

Figure 2. NNSA's Classified Supercomputing Network Infrastructure.

NNSA reported obligating approximately $1.7 billion from fiscal years 2007 through 2009 to support ASC program activities at the three weapons laboratories.[7] The $1.7 billion was predominantly associated with three efforts:

Weapons codes and models. This effort is intended to develop and improve weapons simulation models and codes for predicting the behavior of weapons systems and devices in the nuclear stockpile.

Computational systems and software environment. This effort is intended to provide ASC users a stable, seamless computing environment for ASC-deployed platforms. It is responsible for procuring, delivering, and deploying ASC computational systems and user environments via technology development and integration across the three weapons laboratories.

Facility operations and user support. This effort is intended to provide both the necessary physical facility and operational support for reliable supercomputing and storage environments, as well as a suite of user services for effective use of the three weapons laboratories' computing resources. Facility operations cover physical space, power and other utility infrastructure, and local- and wide-area networking, as well as system administration, cyber security, and operations services for ongoing support. The user support function includes planning, development, integration and deployment, continuing product support, and quality and reliability activity collaborations.

To strengthen the security of information and information systems across the federal government, including those at NNSA's weapons laboratories, the *Federal Information Security Management Act of 2002* (FISMA) requires each agency to develop, document, and implement an agencywide information security program that supports the operations and assets of the agency, including those provided or managed by another agency or contractor on its behalf.[8] This security program is to include plans and procedures to ensure the continuity of operations for information systems that support the agency's operations.[9] Pursuant to its FISMA responsibilities, the National Institute of Standards and Technology (NIST) has issued federal standards and guidelines on information security, such as a contingency planning guide for federal information systems, and recommended security controls, which address contingency and disaster recovery planning and testing.[10] To further ensure the security of national security systems, the Committee on National Security Systems (CNSS)[11] requires federal agencies with national security systems to implement a comprehensive set of security controls and enhancements for these systems.[12] CNSS requires that each agency implement a contingency and disaster recovery planning capability that ensures the integrity and availability of its national security information and information systems.[13]

FISMA, NIST guidelines,[14] and CNSS policies all call for contingency and disaster recovery planning—also referred to as continuity of operations for information systems—for critical components of information protection. DOE and NNSA policies also regard contingency and disaster recovery plans as being necessary for information protection. If normal operations are interrupted, contingency and disaster recovery plans allow senior agency officials to detect, mitigate, and recover operations. Examples of the key components that make up contingency and disaster recovery planning programs include (1) assessing the criticality and sensitivity of computerized operations and identification of supporting resources such as developing business impact analyses (BIA), (2) taking steps to prevent and minimize potential damage and interruption such as establishing data backup processes, (3) developing comprehensive contingency and disaster recovery plans,[15] and (4) conducting periodic testing of contingency and disaster plans.

The extent to which controls—such as contingency and disaster recovery planning—are implemented depends on a level of risk assigned to the system or information maintained on the system. NIST standards and guidelines,

CNSS instructions, and NNSA policy allow consideration of risk in determining the level of protection of systems and data. These standards and policies require that organizations consider the impact or consequences of loss as it relates to the confidentiality, integrity, and availability of the information, and assign a value of low, moderate, or high impact levels. For contingency and disaster recovery planning, consideration of "availability" is the key element. NNSA policy defines the values for the consequences of loss associated with availability as follows:

High—Loss of life might result from loss of availability; information must always be available on request, with no tolerance for delay; loss of availability will have an adverse effect on national-level interests; federal requirement (i.e., requirement for material control and accountability inventory); or loss of availability will have an adverse effect on confidentiality.

Moderate—Information must be readily available with minimum tolerance for delay; bodily injury might result from loss of availability; or loss of availability will have an adverse effect on organizational-level interests.

Low—Information must be available with flexible tolerance for delay.

NNSA Has Not Fully Implemented Contingency and Disaster Recovery Planning and Testing for Its Classified Supercomputing Assets

Contingency and disaster recovery planning and testing for NNSA's classified supercomputing systems have not been fully implemented at each of the three weapons laboratories—Los Alamos, Sandia, and Livermore. Specifically, NNSA did not ensure that the laboratories (1) developed BIAs to determine the impact of potential service disruptions, (2) fully tested data backup processes, and (3) developed and tested contingency and disaster recovery plans. These shortcomings existed, at least in part, because NNSA's component organizations were unclear of their roles and responsibilities for providing oversight in the laboratories' implementation of contingency and disaster recovery planning. Until the agency fully implements a contingency and disaster recovery planning program for its classified supercomputing assets at the weapons laboratories, it has limited assurance that vital information can be recovered and made available to meet national security priorities and requirements.

Not All of the Laboratories Assessed the Criticality and Sensitivity of Supercomputer Operations and Resources, or Potential Outage Impact

To assess the criticality and sensitivity of computerized operations and identification of supporting resources, NIST guidelines state that agencies should determine their recovery strategies by performing business impact analyses of their systems. A BIA is an analysis of information technology system requirements, processes, and interdependencies used to characterize system contingency requirements and priorities in the event of a significant disruption. NIST guidelines state that agencies conduct a BIA to identify critical information systems to fully characterize the system's requirements, processes, and interdependencies to determine contingency requirements and priorities. In addition, according to NIST guidelines, the BIA process should follow three main steps: (1) identify critical data and information technology resources, (2) identify outage impacts and allowable outage times,[16] and (3) develop recovery priorities and strategies. NNSA policy also requires a BIA to identify systems that provide critical services to site operations and prioritize these systems and their components.

One of the laboratories—Los Alamos—had conducted a BIA that addressed its classified supercomputing systems, generally following the three steps of a BIA. However, the BIA was not always specific. For example, the laboratory identified critical information technology resources for each of its classified supercomputing systems, but did not specifically identify the critical data. Instead, Los Alamos noted that the systems are not considered mission critical nor mission essential to the business needs of the laboratory, [17] and that the consequence of loss for system availability is low. Additionally, it defined a specific number of days for the allowable time frames for fully and partially disabled systems, but did not provide specifics on allowable outage impacts. Further, the analyses indicated high-level recovery priorities, but did not provide specifics regarding the recovery process or strategies that would be used for recovery efforts.

The other two laboratories did not conduct BIAs specifically for classified supercomputing systems, but plan to do so. Livermore has a BIA in place for its logical assets—the applications and services that provide basic operational support to the Livermore computing environment, but the BIA did not address any of the classified supercomputing systems. However, at the time of our site visit, Livermore officials stated they were beginning the process of developing a BIA that would address their information technology needs for their classified supercomputing systems, but the process was still in the planning stage. Similarly, according to Sandia officials, the laboratory has BIAs that address its unclassified information technology systems, but does not currently have one specifically for its classified supercomputing systems. However, Sandia officials indicated that they plan to conduct a BIA for classified supercomputing systems in 2011.

Although the two laboratories have not conducted any BIAs—in line with the BIA conducted by Los Alamos—they have considered the risk of consequence of loss from availability as low impact. NNSA also considers the consequence of loss as low impact. In addition, NNSA and the three laboratories do not consider the classified supercomputers to be "mission critical." One laboratory categorized the systems as "mission essential," while another referred to them as "mission support elements, not mission essential elements." However, NNSA's mission includes maintaining the safety, security, and effectiveness of the nuclear deterrent without nuclear testing. The supercomputers provide a necessary means to determine the effects of changes to current weapons systems and to determine a level of confidence in the performance of future untested systems. The classified supercomputing capabilities serve as the computational surrogate to nuclear weapons testing and are central to national security.

Regarding recovery priorities and strategies, each of the laboratories indicated that it would likely rely on a process that is currently being used for the capability system shared among the laboratories. The laboratories generally rely on the Capability Computing Campaign to prioritize the workload and develop priorities for jobs that need to be run on the capability system.[18] In the event of a service disruption or emergency, laboratory officials told us that they would likely rely on the same process for all of their systems. However, this process has not been documented as a means for establishing overall recovery priorities across the laboratories.

Until all of the laboratories have a BIA in place for their classified supercomputing systems that (1) identifies and categorizes critical data, identifies acceptable allowable outage impacts and time frames, and establishes emergency processing priorities and strategies, the potential impact of a system outage will remain unclear.

The Laboratories Have Backup Processes in Place, but One Storage Site May Be Susceptible to Damage

Data backup processes offer a means of taking steps to prevent and minimize potential damage and interruption to computerized services. NIST guidelines, as well as CNSS instructions and NNSA policies, call for agencies to conduct backups of user-level information, system-level information, and information system documentation. In addition, NIST, CNSS, and NNSA all provide that agencies establish an alternate storage site that is separated from the primary storage site so that both are not susceptible to the same hazards. To ensure the availability of data stored in the alternate storage site, NIST and CNSS require that agencies test the backup information to verify the integrity of the data.

All of the laboratories had backup processes in place. Each of the laboratories follows similar data backup processing—both manual and automated procedures—to back up user-level information, system-level information, and information system documentation. For example, this information can include global directories, user home directories, project directories, desktop systems, and critical systems documentation. Backups occur in increments: daily incremental backups to disk, weekly full backups to tape, and monthly full-system backups to tape (with a 6- month on-site storage retention policy). The laboratories also have vendor- provided software that takes periodic snapshots of user directories for storage retention purposes. The snapshot process can be performed manually or can be set up for automatic processing. Users are encouraged to maintain their data in a shared environment on the network and are allowed to make their own determinations regarding what data should be backed up from the classified supercomputing systems.

Not all of the laboratories have an alternate storage site sufficiently separated from the primary site to not be susceptible to the same hazards. Two of the three laboratories have alternate storage sites a considerable distance from their primary storage site. Livermore sends its system backups electronically to Los Alamos every 6 months. Sandia sends its backup data to its alternate site locations (e.g, the California site sends its data to the New Mexico site and the New Mexico site sends its data to the California site). However, Los Alamos maintains its alternate storage facility on-site in a building located less than 1 mile away from the primary local backup storage facility. Consequently, both sites could be susceptible to the same hazards, such as a wildfire.

The laboratories had processes in place to verify the integrity of the backed up data. However, tests of their backup procedures rely predominantly on ad hoc recovery, rather than periodically planned tests. Los Alamos officials indicated that thousands of file recoveries have been performed over the years by end users as part of their testing. Livermore officials stated that the laboratory tests its local backup procedures through actual system usage on almost a daily basis, and tests their remote backup procedures at least once annually. Further, Sandia officials told us they had successfully tested a sample of data at their offsite facility.

Not All Laboratories Had Developed and Tested Contingency and Disaster Recovery Plans

NIST guidelines and CNSS policies call for the development and testing of contingency plans and the development of disaster recovery plans for each information system to ensure that, in the event of a service disruption, the work and supporting functions of the agency can continue to be performed. According to NIST guidelines, at a minimum, the contingency plan should address the identification and notification of key personnel, plan activation, system recovery, and system reconstitution to meet the needs of the agency's critical supporting operations. The guidelines also state that the plan should be tested periodically; CNSS specifies that the frequency of testing should be annually. NIST also notes that the disaster recovery plan should be designed to restore operability of the targeted system, application, or computer facility at an alternate site after a major service disruption. DOE and NNSA policies also require the development of contingency and disaster recovery plans and the testing of these plans in line with NIST and CNSS.

Each of the laboratories had developed contingency plans for their classified supercomputing systems; however, the plans were not always comprehensive, and at the time of our site visits, these plans had not been tested. The laboratories addressed disaster recovery planning to a limited extent; none specifically addressed the supercomputing environment. For example,

- Two laboratories—Los Alamos and Sandia—had contingency plans that addressed the classified supercomputing systems. Although Livermore had an information technology contingency plan and a master security plan, neither specifically addressed the supercomputers. In addition, the plans for both Los Alamos and Sandia included key components such as the identification and notification of key personnel, plan activation, system recovery, and system reconstitution procedures; however, the sufficiency of the level of detail varied. For instance, one plan provided specific details regarding system recovery processes and the notification and identification of key personnel, but provided limited details regarding plan activation and system reconstitution procedures.

- At the time of our site visits, none of the laboratories had tested their contingency plans, which were less than a year old. One of the laboratories—Los Alamos—had created testing guides but had not yet conducted formal testing. Subsequent to our site visit, Los Alamos officials indicated that the first test of their plan took place in September 2010 and noted that the results would be finalized in December. Additionally, although Sandia had a contingency plan in place, the plan states that testing is not required because, in the event of a service disruption, the laboratory would either wait until the equipment was fully operational or simply acquire new equipment. This is contrary to NIST guidelines, CNSS instructions, and DOE and NNSA policies.
- Each of the laboratories had addressed disaster recovery planning to a limited extent. For example,
  - Los Alamos included disaster recovery planning as a section within their classified supercomputing system's contingency plans. Although it provided high-level instructions such as directing individuals to call 911 for all emergencies, it did not include information regarding the specifics for restoring operability of the classified supercomputing system at an alternate site after a major service disruption.
  - None of the plans submitted by Livermore specifically addressed the supercomputing environment, although in the disaster recovery section of the master security plan, the laboratory noted that it had no mission-essential systems in the computing environment, and systems may be offline for an extended period for system upgrades.
  - Sandia also had disaster recovery plans in place, designed for emergency preparedness and disease response planning needs for the laboratory. These plans focused on emergencies involving the facilities, operations, and activities for the laboratory, and

provided individuals with emergency information should pandemics plague the laboratory. However, these plans did not include any information regarding the classified supercomputing systems.

Unless each of the laboratories develops and sufficiently tests comprehensive contingency and disaster recovery plans in accordance with applicable policies and guidance for their classified supercomputing systems, they face a risk of not being able to successfully recover their supercomputing assets and operations after a service disruption.

NNSA Component Organizations Were Unclear of Their Roles and Responsibilities for Providing Oversight

The aforementioned shortcomings existed, at least in part, because NNSA's component organizations were unclear of their roles and responsibilities for providing oversight in the laboratories' implementation of contingency and disaster recovery planning. FISMA requires that the chief information officer, in coordination with other senior agency officials, manage the development and implementation of an agency wide information security program that includes plans and procedures to ensure continuity of operations for information systems that support the operations and assets of the agency. NIST guidelines and DOE policies call for individuals with information system or security management and oversight responsibilities to take responsibility for the development, implementation, assessment, monitoring, reviewing, and updating of security planning policies and procedures, which includes contingency and disaster recovery plans. Further, the *NNSA Safety Management Function and Responsibilities and Authorities Manual* states that the chief information officer is responsible for information technology programs and initiatives and for ensuring the security of the agency's information and systems.

Although roles and responsibilities are defined at a high level in FISMA, NIST guidelines, as well as DOE and NNSA policies, NNSA component organizations were confused about their roles in providing oversight of the laboratories' implementation of contingency and disaster recovery planning for the supercomputing systems. For example, at the beginning of our review, ASC officials told us that, although they were responsible for administering and managing the program that uses the classified supercomputing systems, they were not responsible for contingency and disaster recovery planning. Instead, they directed us to the Office of the Chief Information Officer (OCIO), where officials told us that they were not responsible for contingency and disaster recovery planning for these systems, and noted that they would only provide guidance if requested by ASC. Further, OCIO officials told us that ASC has not requested any assistance. ASC officials subsequently acknowledged that they had responsibility for contingency and disaster recovery planning; however, this organizational responsibility is contrary to NIST guidelines and DOE policies, as well as NNSA's own manual, which gives this responsibility to the OCIO. In the absence of effective oversight, the laboratories did not consistently comply with, or fully implement, federal requirements and guidance related to contingency planning and disaster recovery. Until NNSA clearly establishes and carries out defined roles and responsibilities for OCIO and ASC pertaining to contingency and disaster recovery planning for the classified supercomputing environment, it will not be able to effectively manage and oversee the recovery of its supercomputing operations should service disruptions occur.

The Laboratories Have the Ability to Share Supercomputing Capacity, but Barriers Exist

Technologically, the weapons laboratories have demonstrated the ability to share classified supercomputing capacity using their capacity and capability systems under normal operating conditions. Although these supercomputers process unique workloads and operate independently, they are designed with a similar operating system, resource manager, and job scheduler, which is built on a LINUX foundation. These supercomputers also include application codes that are portable across supercomputing systems and a data network, which allows authorized users local and remote access to the systems.

Although the weapons laboratories have the ability to share supercomputing capacity, barriers exist. One barrier to sharing supercomputing capacity is that the weapons laboratories do not know the minimum supercomputing capacity needed to achieve processing priorities in the event of a service disruption. NIST guidelines recommend, and NNSA policy requires, that capacity planning be conducted so that there is adequate capacity for information processing and supporting resources during contingency operations. Although the weapons laboratories have identified the supercomputing processing needed for normal business operations, they have not identified the minimum supercomputing capacity needed to achieve processing priorities in the event of a service disruption.

Another barrier to sharing supercomputing processing is the disparity in usable supercomputing processing across the laboratories. Figure 3 depicts this disparity by identifying the amount of total usable supercomputing capacity, in teraFLOPS, for each of the three weapons laboratories for 2010 and 2011.[19]

For example, in 2010, total usable capacity at Livermore has been 311 teraFLOPS, whereas Los Alamos and Sandia have had 127 teraFLOPS and 76 teraFLOPS, respectively. Should Livermore experience service disruptions for a sustained amount of time, neither Los Alamos nor Sandia possesses the necessary usable supercomputing capacity to accommodate the additional workload and NNSA will have to reprioritize the computational workloads across the other two laboratories. As previously noted, officials at the laboratories told us that, should disruptions occur, they would use the Capability Computing Campaign model for re-prioritizing the workload. However, this process has not been documented for recovery activities.

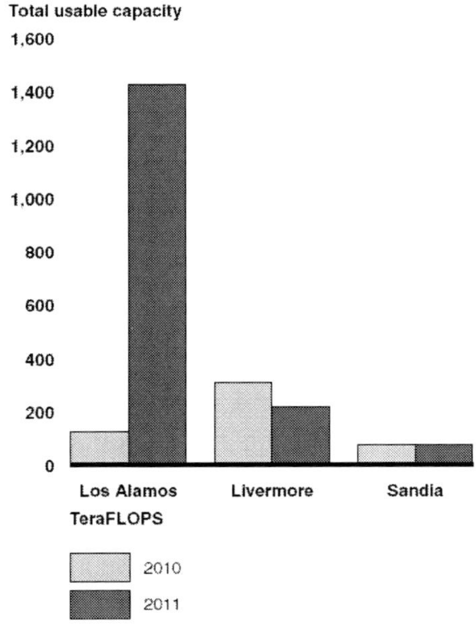

Source: GAO analysis of supercomputing capacity data provided by Los Alamos, Livermore, and Sandia.

Figure 3. Total Usable Supercomputing Capacity at Each Weapons Laboratory, 2010 and 2011.

Further limiting the ability of the weapons laboratories to recover from a service disruption, in 2011, there will be a significant disparity in projected usable supercomputing capacity. For example, in 2011, Los Alamos' usable capacity is projected to be 1,427 teraFLOPS, whereas usable capacity at Livermore and Sandia is to be 218 teraFLOPS and 76 teraFLOPS, respectively. Should Los Alamos' supercomputing systems become unavailable for an extended period of time, neither Livermore nor Sandia possesses sufficient usable supercomputing capacity to achieve its workload and accommodate the additional potential computational workload from Los Alamos. According to laboratory officials, an additional supercomputer will be deployed at Los Alamos in 2011. This supercomputer is a replacement for the single capability supercomputer currently at Livermore that was retired in 2010. Therefore, a significant amount of usable supercomputing capacity will be centralized at Los Alamos. Because the weapons laboratories have not determined the minimum supercomputing capacity requirements for their emergency processing priorities, they may not be able to meet the minimum computation al workload required to meet Stockpile Stewardship milestones.

Another barrier to sharing supercomputing capacity across the weap ons laboratories is that the capability to share usable capacity on an "ondemand"[20] basis has not been fully tested in a recovery scenario. According to officials at the laboratories, during normal operating conditions, simulation programs have run on other supercomputing systems. However, consideration has not been given to include and test these abilities in a disaster recovery scenario should a service disruption occur. As a result, NNSA has limited assurance that its disaster recovery approach would work effectively should a service disruption occur.

NNSA Does Not Track the Costs for Ensuring Contingency and Disaster Recovery Planning for Its Supercomputing Assets

Although NNSA reported obligating approximately $1.7 billion from fiscal 2007 through 2009 to implement its ASC program activities at the three weapons laboratories, the costs for ensuring the recovery of its classified supercomputing operations are unknown. Under GAO's *Standards for Internal Control in the Federal Government*, financial information should be recorded and communicated to program managers who need this information to make operational decisions and to effectively allocate resources for program activities.

NNSA officials reported obligating approximately $390 million for facility operations and user support activities, which include the funds associated with contingency and disaster recovery planning activities, but they were unable to provide detailed financial information for contingency and disaster recovery planning activities. According to NNSA officials, costs for contingency and disaster recovery planning for classified supercomputing systems are unknown because ASC program expenditures are part of the NNSA ASC operational budget, whose costs are tracked at an aggregate level. As a result, neither NNSA nor the three weapons laboratories can track what has been spent since fiscal year 2007 for ensuring the recovery of classified supercomputing operations and, consequently, they do not know whether funding levels for these activities have been adequate. Although certain components of contingency and disaster recovery planning are in place at the three weapons laboratories, NNSA is uncertain as to what funds were spent on these information protection activities.

Furthermore, NNSA has not developed contingency planning and disaster recovery cost estimates for its classified supercomputing assets. For fiscal years 2011 through 2014, NNSA projects that it needs about $2.2 billion to implement its ASC activities, which support its Stockpile Stewardship program—$984 million for weapons codes and models, $604 million for computational systems and software environment, and $588 million for facility operations and user support. Although NNSA has developed its out-year funding needs over the next 4 years, it has not developed estimates regarding the future costs for ensuring the recovery of its classified supercomputing assets in the event of service disruptions. Until NNSA develops a means for tracking current contingency and disaster recovery costs and for developing estimates of future costs, the agency will not have the information needed to determine whether it is meeting its goals for effective stewardship of public resources.

CONCLUSIONS

All three NNSA weapons laboratories have implemented some components of a contingency planning and disaster recovery program. NNSA, however, has not provided effective oversight to ensure that the laboratories have comprehensive and effective contingency and disaster recovery planning and testing. For example, B IAs that identify critical resources and outage impacts have not been developed for all classified supercomputing systems and existing contingency plans at the laboratories have not been thoroughly tested. Although one laboratory's analysis is not comprehensive and the other two laboratories have not completed a BIA, NNSA and the laboratories consider the consequence of loss of availability of the classified supercomputers as a low-risk impact, and do not consider them to be mission critical. However, it is unclear how NNSA made this determination given that (1) the analyses have not been completed; (2) NNSA's mission includes maintaining the safety, security, and effectiveness of the nuclear deterrent without nuclear testing; (3) the classified supercomputing capabilities serve as the computational surrogate to underground nuclear weapons testing and are central to our national security; and (4) NNSA has obligated about $1.7 billion over 3 fiscal years to support the Advanced Simulation and Computing program, which includes classified supercomputing activities.

Beyond the activities undertaken by the laboratories, NNSA has not developed a means for identifying, tracking, or re-prioritizing the classified supercomputing workload across the operating environment. In addition, the laboratories have not tested offsite recovery capabilities and the agency has not tested the laboratories' ability to share "on-demand" capacity if needed or determined the minimum capacity needed to meet Stockpile Stewardship Program requirements, particularly in the event that it may need to establish emergency processing priorities. Further, although over a billion dollars have been obligated to support the classified supercomputing capabilities within the last 3 years, NNSA has not tracked the costs for ensuring the recovery of the classified supercomputing systems, data, and supporting resources should a service disruption occur. The classified supercomputing program represents a significant investment, and accountability for these systems is essential. Until NNSA clearly defines its component organizations' roles and responsibilities and fully implements an effective contingency and disaster planning program, it has limited assurance that, in the event of a service disruption, vital information could be recovered and made available to meet national security priorities.

RECOMMENDATIONS FOR EXECUTIVE ACTION

To improve the effectiveness of contingency and disaster recovery planning for NNSA's classified supercomputing capabilities, we recommend that the Administrator of NNSA direct the weapons laboratories to take the following four actions, where not already implemented:

- Develop business impact analyses that, among other things, (1) identify and prioritize critical systems, data, and supporting resources; (2) identify allowable outage times and impacts for classified supercomputi ng c apabilities; and (3) identify recovery priorities and strategies.
- Develop and implement comprehensive contingency and disaster recovery plans for all classified supercomputing systems that identify how each weapons laboratory's classified supercomputing capabilities will be recovered following service disruptions.
- Conduct contingency and disaster recovery plan testing.
- Test the three weapons laboratories' ability to share "on-demand" classified supercomputing capacity to ensure this capability will work in the event of unexpected service disruptions.

In addition, we recommend that the Administrator of NNSA take the following five actions:

- Document an agencywide means for reprioritizing the workload across NNSA's classified supercomputing systems should a disruption occur.
- Clearly define the oversight responsibilities of the NNSA ASC program office and the NNSA Office of the Chief Information Officer, as they relate to contingency and disaster recovery planning for NNSA's classified supercomputing operations.
- Identify, assess, and communicate the minimum classified supercomputing capacity needed to meet Stockpile Stewardship requirements in the event of a service disruption.
- Develop, document, and implement a process that identifies and tracks expenditures for contingency and disaster recovery planning for NNSA's classified supercomputing assets.
- Develop and document the total anticipated costs for contingency and disaster recovery planning of NNSA's classified supercomputing assets, which includes the replacement costs for these assets.

AGENCY COMMENTS AND OUR EVALUATION

In providing written comments on a draft of this report, NNSA's Associate Administrator for Management and Administration agreed that improvements can be made in contingency and disaster recovery planning for supercomputing operations. He indicated that NNSA agreed with six of our nine recommendations and outlined the agency's intent to conduct business impact analyses, develop and test appropriate contingency and disaster recovery plans, document workload prioritization, and clearly define roles and responsibilities.

However, NNSA did not agree with our recommendation related to identifying the minimum capacity needed to meet Stockpile Stewardship requirements in the event of a service disruption. The Associate Administrator stated that this recommendation did not take into account that the different types of supercomputers—capacity and capability— serve different functions and are procured and managed differently. In our report, we recognize that different types of supercomputers exist and that they are used for different purposes, they process unique workloads and operate independently. However, as we point out in the report, although the weapons laboratories have identified supercomputing processing needed for normal business operations, they have not identified the minimum capacity needed to achieve processing priorities in the event of a service disruption. We believe that the recommendation appropriately focuses on meeting NNSA's Stockpile Stewardship mission and that capacity planning is essential to ensure that information processing and supporting resources exist during contingency operations, regardless of the type of system used. Although NNSA did not agree with the recommendation, the Associate Administrator stated that the agency will conduct a BIA and build appropriate contingency strategies for both types of supercomputers, as well as enhance capacity sizing actions to account for contingency and disaster recovery operations.

Further, NNSA did not agree with two recommendations related to identifying and tracking expenditures for contingency and disaster recovery planning and documenting anticipated recovery planning costs , including replacement costs of the assets. The Associate Administrator asserted that this information would not add significant value to managing contingency and disaster recovery planning. However, we believe such actions reflect good government practices and would add value by providing NNSA program managers with useful expenditure and cost information to aid decision making with regards to contingency and disaster recovery planning. As our report points out, GAO's *Standards for Internal Control in the Federal Government* states that financial information should be recorded and communicated to program managers to help them make operational decisions and effectively allocate resources for program activities. Strong financial and internal controls are a major part of managing any organization because they help government program managers achieve desired results through effective stewardship of public resources. Accordingly, we believe our recommendations have merit.

Gene Aloise
Director, Natural Resources and Environment

Nabajyoti Barkakoti
Director, Center for Technology and Engineering

Gregory C. Wilshusen
Director, Information Security Issues

APPENDIX I. OBJECTIVES, SCOPE AND METHODOLOGY

The objectives of our review were to assess the extent to which (1) the National Nuclear Security Administration (NNSA) has implemented contingency and disaster recovery planning and testing for its classified supercomputing assets, (2) the three laboratories are able to share classified supercomputing capacity for recovery operations, should service disruptions occur, and (3) NNSA tracks the costs for ensuring contingency and disaster recovery planning for its classified supercomputing assets. To address these objectives, we focused on contingency and disaster recovery planning activities at NNSA headquarters, as well as the operating environment for the 12 classified supercomputing systems at the three weapons laboratories—Los Alamos National Laboratory, Livermore National Laboratory, and Sandia National Laboratories.

To assess the extent to which NNSA has implemented contingency and disaster recovery planning and testing for its classified supercomputing assets, we examined contingency and disaster recovery planning controls for the systems within the classified supercomputing environment that are critical to NNSA's achievement of its nuclear weapons mission. We collected and reviewed policies, procedures, and guidelines from the National Institute of Standards and Technology, the Committee on National Security Systems, the

Department of Energy, and NNSA. We also reviewed contingency plans and business impact analyses provided by the weapons laboratories and compared them to federal guidelines. We interviewed NNSA and laboratory officials to determine whether they had documented critical system, data, and supporting resources and whether contingency plans had been tested. Further, we interviewed NNSA officials to determine to what extent they have provided specific guidance and oversight for the laboratories to ensure that contingency and disaster recovery planning requirements are being met.

To determine the extent to which the three weapons laboratories have the ability to share supercomputing capacity for backup and recovery operations, we visited each weapons laboratory and gained an understanding of the overall classified supercomputing infrastructure and identified interconnectivity and control points. We performed technical assessments of supercomputing capabilities at each weapons laboratory, including each laboratory's ability to share supercomputing capacity under normal operating conditions. We reviewed the weapons laboratories' efforts to determine the minimal supercomputing capacity needed to meet NNSA Stockpile Stewardship Program requirements along with the ability of the weapons laboratories to share supercomputing capacity on an "on- demand" basis, including the use of advanced architecture systems. In addition, we obtained documents describing the supercomputing system environment as well as capacity information, along with the views of officials from NNSA and the three weapons laboratories.

To assess the extent to which NNSA tracks costs for ensuring contingency and disaster recovery planning for classified supercomputing assets, we interviewed NNSA and weapons laboratory officials to determine how expenditures were tracked for contingency and disaster recovery planning of the classified supercomputing systems at each of the laboratories. We also requested the amount of funds NNSA obligated to the three weapons laboratories, and the amount of funds the laboratories spent, in implementing NNSA's classified supercomputing capabilities from fiscal years 2007 through 2009. Further, we interviewed NNSA and laboratory officials to determine how they projected future cost estimates for ensuring the recovery of these assets for fiscal years 2011 through 2014. To assess the reliability of data provided, we reviewed (1) the fiscal year 2009 financial statement audit for the system and (2) responses NNSA provided to questions about processes and procedures for ensuring the accuracy and completeness of data. Based on this information, we determined the data are sufficiently reliable for the purposes of this report.

We conducted this performance audit from December 2009 through December 2010 in accordance with generally accepted government auditing standards. Those standards require that we plan and perform the audit to obtain sufficient, appropriate evidence to provide a reasonable basis for our findings and conclusions based on our audit objectives. We believe that the evidence obtained provides a reasonable basis for our findings and conclusions based on our audit objectives.

APPENDIX II. NNSA ANNUAL OBLIGATIONS FOR ITS ADVANCED SIMULATION AND COMPUTING PROGRAM, FISCAL YEARS 2007 THROUGH 2009

NNSA reported obligating approximately $1.7 billion from fiscal years 2007 through 2009 to support Advanced Simulation and Computing (ASC) program activities at the three weapons laboratories. The $1.7 billion was used mainly for three efforts:

Weapons codes and models. This effort is intended to develop and improve weapons simulation codes and models for predicting the behavior of weapons systems and devices in the nuclear stockpile.

Computational systems and software environment. This effort is intended to provide ASC users a stable, seamless computing environment for ASC-deployed platforms. It is responsible for procuring, delivering, and deploying ASC computational systems and user environments via technology development and integration across the three weapons laboratories.

Facility operations and user support. This effort is intended to provide both the necessary physical facility and operational support for reliable supercomputing and storage environments, as well as a suite of user services for effective use of the three weapons laboratories' computing resources. Facility operations cover physical space, power and other utility infrastructure, and local- and wide-area networking, as well as system administration, cyber security, and operations services for ongoing support. The user support function includes planning, development, integration and deployment, continuing product support, and quality and reliability activity collaborations.

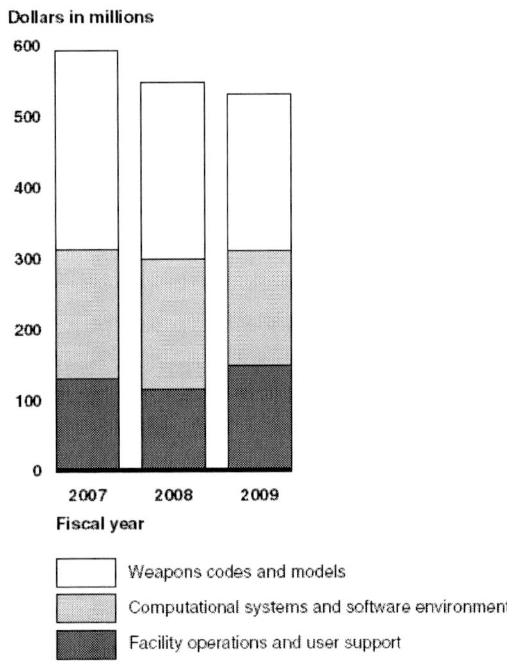

Source: GAO analysis of data provided by NNSA.

Figure 4. Annual Obligations for NNSA's Advanced Simulation and Computing Program, Fiscal Years 2007 through 2009.

Figure 4 depicts NNSA's annual obligations for each of the three efforts from fiscal years 2007 through 2009.

As shown in figure 4, NNSA annual obligations for its classified supercomputing operations decreased from about $591 million to $529 million between fiscal years 2007 and 2009. The largest obligation for the classified supercomputing program was for weapons codes and models, which accounted for approximately $750 million (or 45 percent) of total obligations.

Obligations for computational systems and software environment accounted for approximately $527 million (or 32 percent) of total obligations. For the period, obligations for this effort decreased from $183 million to $161 million.

The facility operations and user support activities, which includes, among other things, expenditures for contingency and disaster recovery planning, accounted for $390 million (or 23 percent) of total obligations over the period.

These obligations ranged from $114 million to $147 million for the 3 fiscal years.

End Notes

[1] NNSA was established in 2000 as a separately organized agency within the Department of Energy (DOE) and is responsible for the nation's nuclear weapons, nonproliferation, and naval reactors programs.

[2] For nearly half a century, the United States' nuclear program was spearheaded by underground nuclear testing and never had to rely on weapon systems that had exceeded their design life times. The United States last produced a nuclear weapon in 1991 and performed its last underground nuclear test in 1992.

[3] The *National Defense Authorization Act for Fiscal Year 1994*, Pub. L. No. 103-160, § 3138 (1993), directed DOE to establish the Stockpile Stewardship Program. In the absence of underground nuclear testing, the program encompasses a broad range of activities to increase understanding of the basic phenomena associated with nuclear weapons, provide better predictive understanding of the safety and reliability of weapons, and ensure a strong scientific and technical basis for future nuclear weapons policy objectives. The Stockpile Stewardship Program is carried out through the nuclear weapons complex, which includes three nuclear weapons laboratories.

[4] Continuity of operations focuses on restoring an organization's mission-essential functions at an alternate site and performing those functions for a short period of time before returning to normal operations. Contingency and disaster recovery planning include a broad scope of activities designed to sustain and recover critical information and information system services for a range of potential service disruptions. Contingency and disaster recovery planning components may include the relocation of information systems and operations to an alternate site, recovery of information system functions using alternate equipment, or the performance of information system functions using alternative methods. For the purposes of this report, the term contingency and disaster recovery planning refer to the interim measures NNSA should use to recover information system services after an unexpected service disruption.

[5] Los Alamos is managed and operated by Los Alamos National Security, LLC, which is a consortium of contractors that includes Bechtel National, the University of California, the Babcock and Wilcox Company, and the Washington Division of URS. Livermore is managed and operated by Lawrence Livermore National Security, LLC, which is comprised of a corporate management team that includes Bechtel National, the University of California, the Babcock and Wilcox Company, and the Washington Division of URS. Sandia is managed and operated by Sandia Corporation, a wholly owned subsidiary of Lockheed Martin Corporation.

[6] FLOPS are a measure of a supercomputing system's performance. Floating-point performance is the rate at which a computer executes floating-point operations.

[7] For additional information regarding budgetary information for the classified supercomputing program from fiscal years 2007 through 2009, see appendix II.

[8] 44 U.S.C. § 3544(b); FISMA was enacted as title III, E-Government Act of 2002, Pub. L. No. 107-347, 116 Stat. 2899, 2946 (Dec. 17, 2002).

[9] For the purposes of this report, we will refer to "continuity of operations procedures for information systems" as contingency and disaster recovery planning.

[10] NIST Special Publication 800-34, *Contingency Planning Guide for Information Technology Systems* (Washington, D.C.: June 2002) and NIST Special Publication 800-53 Revision 3,

Recommended Security Controls for Federal Information Systems and Organizations (Gaithersburg, Md.: August 2009).

[11] Formerly known as the National Security Telecommunications and Information Systems Security Committee, CNSS provides a forum for the discussion of policy issues, sets national policy, and provides direction, operational procedures, and guidance for the security of national security systems. DOD chairs the committee under the authorities established by National Security Directive 42, *National Policy for the Security of National Security Telecommunications and Information Systems*, issued in July 1990. This directive designates the Secretary of Defense and the Director of the National Security Agency as the Executive Agent and National Manager, respectively. The committee has 21 voting representatives from various departments and agencies, including the Department of Energy.

[12] National security systems include any information system used or operated by an agency, or by a contractor of an agency, that processes, stores, or transmits national security information. They do not include those systems used for routine administrative and business applications.

[13] CNSS Instruction 1253 provides federal government departments, agencies, bureaus, and offices with a process for security categorization of national security systems that collect, generate, process, store, display, transmit, or receive national security information. In addition, this instruction serves as a companion document to NIST Special Publication 800-53, Revision 3.

[14] Although NIST guidelines note they shall not apply to national security systems without the express approval of appropriate federal officials exercising policy authority over such systems, CNSS instructions, as well as DOE and NNSA policies for national security systems, refer to the NIST guidelines as being applicable.

[15] A contingency plan is designed to maintain or restore business operations, including computer operations, possibly at an alternate location in the event of emergencies, system failures, or disaster. A disaster recovery plan is a written plan for processing critical applications in the event of a major hardware or software failure or destruction of facilities.

[16] Outage impacts and allowable outage times enable the organization to develop and prioritize recovery strategies that personnel will implement during contingency plan activation. The effects of the outage may be tracked over time, which will enable the organization to identify the maximum allowable time that a resource may be unavailable before it inhibits the performance of an essential function. The effects of the outage can also be tracked across related resources, identifying any cascading effects that may occur as an effect of a service disruption.

[17] The Department of Energy defines *"mission critical"* as an information system that supports an organization's core missions and goals, and *"mission-essential (or business essential)"* as an information system whose failure would not preclude organizations from accomplishing core business functions in the long term.

[18] The Capability Computing Campaign includes a committee made up of staff from the NNSA ASC program office, as well as ASC executives located at the laboratories at Los Alamos, Livermore, and Sandia.

[19] Total usable supercomputing capacity includes the supercomputers that have the ability to run all weapons program codes and could be used in the event of a service disruption, and includes capacity and capability systems.

[20] The term "on demand" is defined as the ability to move an application (simulation program/code) from one supercomputer to a different supercomputer at a different physical facility and use the existing computational resources without the need for major modifications.

In: Federal Information Security
Editors: D. G. Miller and I. I. Wilson
ISBN 978-1-61942-771-6

Chapter 4

INFORMATION SECURITY: CONCERTED EFFORT NEEDED TO CONSOLIDATE AND SECURE INTERNET CONNECTIONS AT FEDERAL AGENCIES

United States Government Accountability Office

WHY GAO DID THIS STUDY

To reduce the threat to federal systems and operations posed by cyber attacks on the United States, the Office of Management and Budget (OMB) launched, in November 2007, the Trusted Internet Connections (TIC) initiative, and later, in 2008, the Department of Homeland Security's (DHS) National Cybersecurity Protection System (NCPS), operationally known as Einstein, became mandatory for federal agencies as part of TIC. For each of these initiatives, GAO was asked to (1) identify their goals, objectives, and requirements; (2) determine the status of actions federal agencies have taken, or plan to take, to implement the initiatives; and (3) identify any benefits, challenges, and lessons learned. To do this, GAO reviewed plans, reports, and other documents at 23 major executive branch agencies, interviewed officials, and reviewed OMB and DHS guidance.

WHAT GAO RECOMMENDS

GAO is making recommendations to OMB to promptly communicate the number of approved connections for agencies, and to DHS aimed at improving communication and performance measures. OMB concurred with GAO's findings, conclusions, and recommendations. DHS concurred with GAO's recommendations and also provided technical comments.

WHAT GAO FOUND

The goals of TIC are to secure federal agencies' external network connections, including Internet connections, and improve the government's incident response capability by reducing the number of agencies' external network connections and implementing security controls over the connections that remain. In implementing TIC, agencies could either provide their own access points by becoming an access provider or seek service from these providers or an approved vendor. To achieve the initiative's goals, agencies were required to

- inventory external connections,
- establish a target number of TIC access points,
- develop and implement plans to reduce their connections,
- implement security capabilities (if they chose to be an access provider) addressing such issues as encryption and physical security, and
- demonstrate to DHS the consolidation of connections and compliance with the security capabilities (if they chose to be an access provider).

As of September 2009, none of the 23 agencies had met all of the requirements of the TIC initiative. Although most agencies reported that they have made progress toward reducing their external connections and implementing critical security capabilities, most agencies have also experienced delays in their implementation efforts. For example, the 16 agencies that chose to become access providers reported that they had reduced their number of external connections from 3,286 to approximately 1,753. Further, agencies have not demonstrated that they have fully implemented the required security capabilities. Throughout their reduction efforts, agencies have experienced benefits, such as improved security and network management. However, they have been challenged in implementing TIC because OMB did not promptly communicate the number of access points for which they had been approved and DHS did not always respond to agency queries on security capabilities in a timely manner. Agencies' experiences with implementing TIC offered OMB and DHS lessons learned, such as the need to define program requirements before establishing deadlines and the usefulness of sponsoring collaborative meetings for agencies' implementation efforts.

Einstein is intended to provide DHS with an increased awareness of activity, including possible security incidents, on federal networks by providing intrusion detection capabilities that allow DHS to monitor and analyze agencies' incoming and outgoing Internet traffic. As of September 2009, fewer than half of the 23 agencies had executed the required agreements with DHS, and Einstein 2 had been deployed to 6 agencies. Agencies that participated in Einstein 1 improved identification of incidents and mitigation of attacks, but DHS will continue to be challenged in understanding whether the initiative is meeting all of its objectives because it lacks performance measures that address how agencies respond to alerts.

ABBREVIATIONS

DHS	Department of Homeland Security
GSA	General Services Administration
NCPS	National Cybersecurity Protection System
OMB	Office of Management and Budget
TIC	Trusted Internet Connections
US-CERT	United States Computer Emergency Readiness Team

March 12, 2010

The Honorable Joseph I. Lieberman
Chairman
The Honorable Susan M. Collins
Ranking Member
Committee on Homeland Security and Governmental Affairs
United States Senate

The Honorable Thomas R. Carper
Chairman
Subcommittee on Federal Financial Management, Government Information, Federal Services, and International Security
Committee on Homeland Security and Governmental Affairs
United States Senate

Pervasive and sustained cyber attacks against the United States continue to pose a potentially devastating impact on federal systems and operations. The need for a vigilant approach to information security is demonstrated by a dramatic increase in reports of security incidents, the wide availability of hacking tools, and steady advances in the sophistication and effectiveness of attack technology. As recently as July 2009, press accounts reported that a widespread and coordinated attack over the course of several days targeted Web sites operated by major government agencies, including the Departments of Homeland Security and Defense, the Federal Aviation Administration, and the Federal Trade Commission, causing disruptions to the public availability of government information. In addition, the Director of National Intelligence testified in February 2009 that foreign nations and criminals had targeted government and private-sector networks to gain a competitive advantage or potentially disrupt or destroy them, and that terrorist groups had expressed a desire to use cyber attacks as a means to target the United States.[1] Such attacks and threats highlight the importance of developing a concerted response to safeguard federal information systems.

To improve the effectiveness of information security across the federal government, in November 2007, the Office of Management and Budget (OMB) announced the Trusted Internet Connections (TIC) initiative, and in 2003 the Department of Homeland Security (DHS) established the Einstein program, recently incorporated into the National Cybersecurity Protection System (NCPS). TIC is intended to improve security by reducing and consolidating external network connections and by providing centralized monitoring at a select group of access providers, while Einstein is an intrusion detection system that provides an automated process for DHS to analyze computer network traffic information from agencies. In January 2008, these programs were incorporated into the Comprehensive National Cybersecurity Initiative.[2]

At your request, we evaluated key elements of the implementation of TIC and Einstein at federal agencies. For each of these initiatives, we (1) identified the goals, objectives, and requirements for the initiatives; (2) determined the status of the actions federal agencies have taken, or plan to take, to implement the initiatives; and (3) identified the benefits, challenges, and lessons learned in implementing the initiatives.

To accomplish our objectives, we examined OMB memorandums and DHS guidance in order to identify program requirements, which we confirmed through interviews with OMB and DHS officials. We obtained and analyzed plans, status reports, and other documents and interviewed officials from 23 of the 24 federal agencies listed in the Chief Financial Officers Act.[3] The Department of Defense was not included in our review because it was not required to implement TIC or Einstein. The initiatives include additional agencies which were not included in our review.

We conducted this performance audit between December 2008 and March 2010 in accordance with generally accepted government auditing standards. Those standards require that we plan and perform the audit to obtain sufficient, appropriate evidence to provide a reasonable basis for our findings and conclusions based on our audit objectives. We believe that the evidence obtained provides a reasonable basis for our findings and conclusions based on our audit objectives. Further details of our objectives, scope, and methodology are included in appendix I.

BACKGROUND

As computer technology has advanced, federal agencies have become dependent on computerized information systems to carry out their operations and to process, maintain, and report essential information. Virtually all federal operations are supported by computer systems and electronic data, and agencies would find it difficult, if not impossible, to carry out their missions, deliver services to the public, and account for their resources without these cyber assets. Information security is thus especially important for federal agencies to ensure the confidentiality, integrity, and availability of their systems and data. Conversely, ineffective information security controls can result in significant risk to a broad array of government operations and assets, as the following examples illustrate:

- Computer resources could be used for unauthorized purposes or to launch attacks on other computer systems.
- Sensitive information, such as personally identifiable information, intellectual property, and proprietary business information could be

inappropriately disclosed, browsed, or copied for purposes of identity theft, espionage, or other types of crime.

- Critical operations, such as those supporting critical infrastructure, national defense, and emergency services, could be disrupted.
- Data could be added, modified, or deleted for purposes of fraud, subterfuge, or disruption.

Due to the growing cyber-based threats to federal systems and critical infrastructure, the persistent nature of information security vulnerabilities, and the associated risks, we continue to designate information security as a governmentwide high-risk issue in our most recent biennial report to Congress,[4] a designation we have made in each report since 1997. In July 2009, we reported[5] that almost all 24 major federal agencies had weaknesses in information security controls and that an underlying reason for these weaknesses is that agencies have not fully implemented their information security programs as required under the Federal Information Security Management Act.[6] As a result, federal systems and sensitive information are at increased risk of unauthorized access and disclosure, modification, or destruction, as well as inadvertent or deliberate disruption of system operations and services.

We have previously reported that federal agencies have experienced security breaches in their networks, potentially allowing sensitive information to be compromised, and systems, operations, and services to be disrupted. These examples illustrate that a broad array of federal information and critical infrastructures are at risk:[7]

- The Department of State experienced a breach on its unclassified network, which daily processes about 750,000 e-mails and instant messages from more than 40,000 employees and contractors at 100 domestic and 260 overseas locations.
- The Nuclear Regulatory Commission confirmed that in January 2003, the Microsoft SQL Server worm known as "Slammer" infected a private computer network at the idled Davis-Besse nuclear power plant in Oak Harbor, Ohio, disabling a safety monitoring system for nearly 5 hours.
- Officials at the Department of Commerce's Bureau of Industry and Security discovered a security breach in July 2006. In investigating this incident, officials were able to review firewall logs for an 8-month period prior to the initial detection of the incident, but were

unable to clearly define the amount of time that perpetrators were inside its computers, or find any evidence to show that data was lost as a result.

Because the threats have persisted and grown, in January 2008 the President issued National Security Presidential Directive 54/Homeland Security Presidential Directive 23, establishing the Comprehensive National Cybersecurity Initiative,[8] a set of projects with the objective of safeguarding federal executive branch government information systems by reducing potential vulnerabilities, protecting against intrusion attempts, and anticipating future threats against the federal government's networks. Under the initiative, DHS is to lead several projects to better secure civilian federal government networks, while other agencies, including OMB, the Department of Defense, the Office of the Director of National Intelligence, and other agencies have key roles in other projects, including monitoring military systems and classified networks, overseeing intelligence community systems and networks, and spearheading advanced technology research and development. The initiative's 12 projects can be grouped into three focus areas:

- *Establishing front lines of defense.* This focus area includes initiatives intended to protect the perimeter of federal networks, such as consolidating connections and deploying intrusion detection and prevention systems.
- *Defend against full spectrum of threats.* This focus area includes activities intended to protect national security and intelligence-related information and systems across the federal government.
- *Shape the future environment.* The initiatives in this area are focused on expansion of cybersecurity education and research and development efforts for future technologies and cybersecurity strategies.

Two primary initiatives under the establishing front lines of defense focus area are TIC and Einstein.

Trusted Internet Connections

In November 2007, OMB announced the TIC initiative.[9] Directed by OMB with assistance from DHS, this effort is intended to improve the federal government's security posture and incident response capability by reducing and consolidating external network connections, including Internet connections, currently in use by the government, and by centrally monitoring the traffic passing through these connections for potentially malicious activity. All federal agencies in the executive branch, except for the Department of Defense, are required to implement the initiative. Although the initiative is intended to secure connections to the Internet, other external connections to potentially unsecured systems must also be routed through an approved TIC access point,[10] even if they do not pass through the Internet.[11]

Agencies may implement TIC by serving as their own access provider or by obtaining services from another source. Agencies may choose one of four service options:

- *Single service:* The agency provides services to its own bureaus and components only.
- *Multi-service:* The agency provides services to its own bureaus and components as well as to other agencies.
- *Seeking service:* The agency obtains services from a multi-service agency or through the Networx program. This program, managed by the General Services Administration (GSA), provides an acquisition vehicle for agencies to procure telecommunication, network, wireless, and information technology security services, including TIC services, from among multiple vendors.
- *Hybrid:* The agency both provides services to its own bureaus and components and obtains additional services from a Networx provider.

Of the 23 agencies in our review, 16 have chosen to be a TIC access provider: specifically, 12 have chosen the single service option, 1 chose the multi-service option, and 3 have chosen the hybrid approach. The remaining seven agencies have chosen to seek service from another access provider.[12]

Einstein

NCPS, operationally known as Einstein,[13] was created in 2003 by the United States Computer Emergency Readiness Team (US-CERT)[14] in order to aid in its ability to help reduce and prevent computer network vulnerabilities across the federal government. The initial version of Einstein provided an automated process for collecting, correlating, and analyzing agencies' computer network traffic information from sensors installed at their Internet connections. The Einstein sensors collected network flow records[15] at participating agencies, which were then analyzed by US-CERT to detect certain types of malicious activity. It then coordinated with the appropriate agencies to mitigate those threats and vulnerabilities. US-CERT also used the information from the sensors to create analyses of cross-governmental trends, offering departments and agencies an aggregate picture of external threats against the federal government's networks. Participation in the program was initially voluntary for federal agencies.

In 2008, DHS developed the current iteration of Einstein—Einstein 2—which incorporated network intrusion detection technology into the capabilities of the initial version of the system. Einstein 2 monitors for specific predefined signatures[16] of known malicious activity at federal agency Internet connections and alerts US-CERT when specific malicious network activity matching the predetermined signatures is detected. According to US-CERT, the signatures are not typically included in commercially available databases of known attack signatures, but are developed by US-CERT to look for specific malicious activity based on previous analysis. In addition, participation in Einstein became mandatory as part of the TIC initiative.

Currently being piloted by DHS, Einstein 3 is intended to be an intrusion prevention system that is to automatically detect and respond appropriately to cyber threats before harm is done. Using signatures developed from critical information about foreign cyber threats as determined by the National Security Agency, the system is to draw on commercial technology and specialized government technology to conduct real-time full packet inspection and threat-based decision making on traffic entering or leaving federal agency networks. It is also intended to support enhanced information sharing by US-CERT with federal agencies by giving DHS the ability to provide agencies with automated alerts of detected network intrusion attempts.

Ultimately, TIC and Einstein are intended to work together to build successive layers of defense mechanisms in the federal government's information technology infrastructures. When Einstein is deployed at a TIC location, it monitors inbound and outbound network traffic. Once TIC is fully implemented across the federal government, all traffic passing between the federal civilian networks and the Internet is to be monitored for malicious activity by US-CERT using Einstein and its supporting processes. Figure 1 illustrates how TIC portals interact with the Einstein sensors and the Internet.

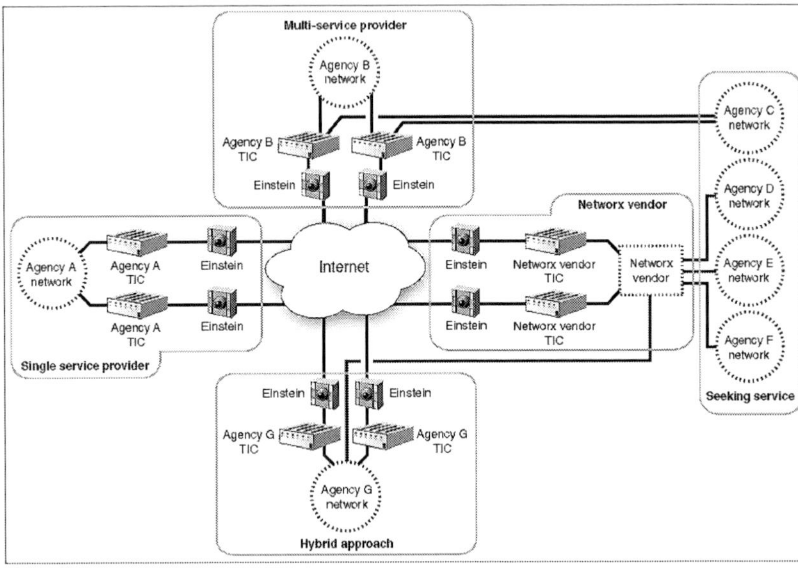

Source: GAO analysis based on DHS data.

Figure 1. Interaction of TIC and Einstein.

AGENCIES HAVE MADE PROGRESS TOWARD CONSOLIDATING AND REDUCING CONNECTIONS, BUT INCONSISTENT COMMUNICATION FROM OMB AND DHS HAS LED TO CHALLENGES

OMB and DHS established requirements to meet the initiative's goals of securing agencies' external connections and improving the government's incident response capability. However, as of September 2009, none of the 23 agencies had met all of the requirements. Throughout their efforts, agencies have experienced benefits and challenges as well as learned lessons.

TIC Aims to Improve the Security of Federal Connections to the Internet

The primary goals of the TIC initiative are (1) to secure federal agency external connections using a common set of security controls and (2) to improve the federal government's incident response capability. To achieve these goals, the initiative has the following objectives:

- reduce and consolidate external connections,[17] including connections to the Internet, across the federal government;
- define and maintain baseline security capabilities for TIC access providers; and
- establish a compliance program to monitor agency adherence to TIC policy.

Agencies Were Required to Develop and Implement Plans to Consolidate and Secure External Connections

To achieve these objectives, agencies were required to:

- **Inventory agency external connections.** Agencies were required to provide their connection inventories to DHS by January 8, 2008.
- **Identify and justify target number of external access points.** Each agency was to submit their target number to DHS by April 15, 2008. They were also required to provide a justification indicating why the requested number of external access points was necessary to support their missions.
- **Develop and implement plans to consolidate external connections.** OMB required agencies to develop and submit initial plans for consolidating their external connections to DHS by January 8, 2008. In addition, agencies were required to update their plans in April 2008. Access provider agencies were required to provide updated plans to DHS in October 2008, and all agencies were required to provide updated plans to DHS in September 2009.
  When it announced the initiative in November 2007, OMB required that agencies' initial plans have a target completion date of June 2008 for reducing and consolidating their external connections. OMB later revised its target deadline for implementation of TIC across the federal government to December 2009.
- **Implement security capabilities.** To ensure that each TIC access point would be secure, OMB required[18] agencies that planned to be an access provider to evaluate their ability to meet 74 security capabilities and to report this information to DHS by April 2008. The 74 security capabilities include technical capabilities, such as encryption of Internet traffic and the use of firewalls; capabilities

related to availability, such as the presence of an uninterrupted power source; physical access controls; and capabilities that describe how an access provider maintains an acceptable level of service. Of the 74 capabilities, 51 are designated as critical, 14 are designated as important, and 9 are categorized as desired. Of the 51 critical capabilities, 40 are required for both single service and multi-service access providers. The 11 capabilities required only for multi-service access providers address the interaction with external customers, such as service level agreements, communication, and reporting.

OMB provided a template for agencies to report whether they currently met each of the capabilities and to indicate their plans for addressing any critical capabilities they did not meet. Once agencies determined whether to be an access provider or to seek service from another provider, they were required to do one of the following:

- Access provider agencies were required to develop plans for implementing any of the critical TIC capabilities that they did not yet have in place. They were required to report on their progress toward implementing the critical capabilities to DHS in October 2008 and September 2009.
- Agencies that are seeking service from other access providers were not required to implement the critical capabilities; however, they were required to acquire TIC services from a multi-service access provider or a commercial vendor that had met the security capabilities through the Networx contract.

- **Demonstrate consolidation of connections and implementation of TIC security capabilities.** Access provider agencies, along with Networx vendors that offer TIC services, are required to undergo a TIC Compliance Validation review, in which DHS assesses the degree to which the access provider meets the critical security capabilities and has consolidated its connections to approved TIC access points. If any capabilities are not fully implemented or if further consolidation is required, the access provider is granted Initial Operating Capability status and is required to develop plans to address the shortcomings and to submit the plans to DHS. All access providers are required to be re-assessed periodically to ensure the capabilities are still being met. All access provider agencies were required to schedule the on-site review with DHS by September 25, 2009.

Agencies Have Not Fully Implemented All Requirements of TIC and Progress Has Been Slower Than Planned

None of the 23 agencies has met all of the requirements of the TIC initiative, and most agencies have experienced delays in their plans for reducing and consolidating connections. However, most agencies reported that they have made progress toward reducing and consolidating their external connections and implementing security capabilities. In addition, several access provider agencies have made more progress toward implementing the capabilities than others. The following describes the status of each requirement.

All Agencies Submitted Connection Inventories

The 23 agencies in our review reported that they initially identified a total of 3,482 external connections. According to DHS, each agency submitted the required inventories, although four submitted the inventories after the January 2008 deadline. Two agencies told us that they discovered additional connections after submitting the initial inventory.

Access Provider Agencies Requested 73 TIC Access Points, but OMB Approved 32

In April 2008, the 16 access provider agencies requested a total of 73 TIC access points. There were a variety of factors that influenced how agencies decided how many access points to request. For example, multiple agencies told us that they chose the number and location of their access points based on the location of existing data centers. Agencies also considered the need for redundant connections, geographic separation between connection sites, the business needs of the agency, and cost factors.

In response to these requests, OMB approved 2 external access points for each access provider agency, a total of 32 TIC access points for the 16 agencies in our review.[19] OMB and DHS established a process for these agencies to request additional access points. As of October 2009, one agency had submitted a request to DHS, and seven other agencies indicated that they had plans to do so.

Progress toward Consolidating Connections Has Been Mixed and Slower than Projected

Progress reported by individual agencies toward meeting their targeted numbers of connections or access points has been mixed, and the reported overall progress toward consolidation has been slower than expected.[20] In submitting their plans, which were due to DHS in October 2008 and September 2009, three agencies reported that they were at their target number of access points and had no further plans to consolidate connections; in addition, one agency did not report the status of its consolidation efforts. Of the remaining 19 agencies, as of September 25, 2009, 6 reported that they had consolidated at least 60 percent of their connections and 9 reported that they had consolidated fewer than 20 percent of their connections. Table 1 shows the consolidation status reported by these 19 agencies as of September 25, 2009.[21]

Overall, the reported progress toward consolidating connections was slower than projected, and agencies delayed their future plans for consolidation. In October 2008, the 16 access provider agencies, which were authorized a total of 32 TIC access points by OMB, projected in their plans of action and milestones that they would consolidate from their initial reported total of 3,286 external connections to a maximum of 1,528 connections by September 2009. However, in their September 2009 plans of action and milestones, these agencies reported that they had consolidated to a maximum of 1,753 connections—225 more than they had planned. In addition, agencies projected in their October 2008 plans that they would have consolidated to a maximum of 764 external connections by OMB's revised deadline of December 31, 2009. However, in September 2009 they anticipated that they would still have a maximum of 1,374 connections by that date—610 more than originally planned—and had significantly revised their projections for consolidation through November 2010. As agencies continue to consolidate their connections, their future projections for consolidation are likely to be revised further. Figure 2 indicates the estimated overall progress that access provider agencies reported toward reducing connections as of October 2008 and September 2009, their planned future consolidation, and how both their plans and reported progress have changed between October 2008 and September 2009.[22]

Table 1. Reported Status of Consolidation by 19 Agencies

Agency type	Reported Status of Consolidation				
	Less than 20%[a]	**20% to 39%**	**40% to 59%**	**60% to 79%**	**80% to 100%**
Access provider	6	1	1	6	0
Seeking service	3	1	1	0	0
Total	**9**	**2**	**2**	**6**	**0**

Source: GAO analysis of agency data.

[a] One access provider agency reported that it was less than 20 percent consolidated on September 25, 2009, but that it expected to consolidate to its target of two connections by September 30, 2009.

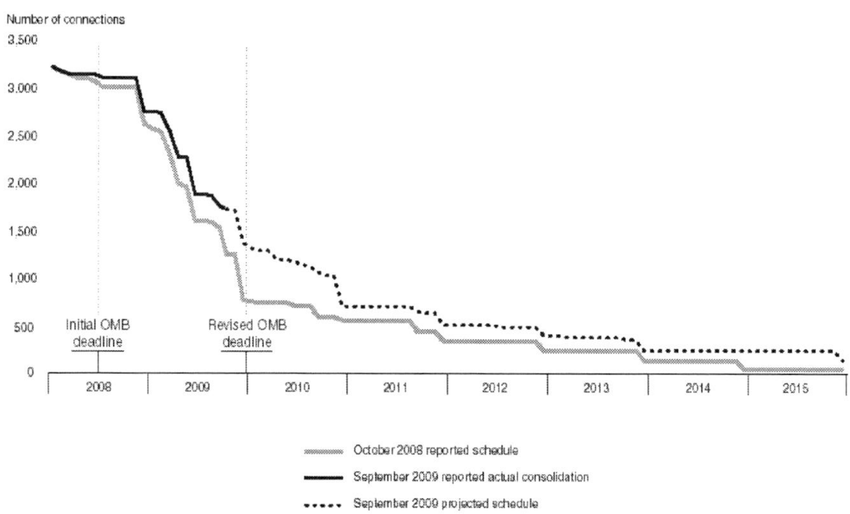

Source: GAO estimate based on agency reported data.

Note: In this figure, both of the reported schedules begin at 3,215 connections because one agency reported that it had consolidated 71 connections by January 2008. In addition, at the time of our review, one access provider agency had not submitted its September 2009 progress report to DHS. As a result, the September 2009 projections for this agency were based on an earlier progress report that may not represent the agency's current status or plans.

Figure 2. Comparison of Reported Consolidation by 16 Access Provider Agencies.

Few Agencies Have Reported Implementing All Required Security Capabilities

As of September 2009, only 3 of the 16 access provider agencies have reported implementing all 40 required critical security capabilities.[23] The other 13 agencies have implemented most of the capabilities, but their progress in addressing the remaining capabilities has varied. For example, of those agencies that had not implemented all of the critical capabilities, six reported meeting no additional capabilities between April 2008 and September 2009. Table 2 describes access provider agencies' reported progress toward implementing the capabilities.

Table 2. Number of Critical Security Capabilities Reported as Implemented by Access Provider Agencies

Agency	Capabilities reported as implemented in April 2008	Capabilities reported as implemented in September 2009	Change between April 2008 and September 2009
A	27	27	0
B	32	32	0
C	33	34	1
D	33	35	2
E	33	36	3
F	34	38	4
G	35	37	2
H	37	37[a]	0
I	37	38	1
J	37	39	2
K	38	38	0
L	38	38	0
M	38	40	2
N	39	39	0
O	40	40	N/A
P	40	40	N/A

Source: GAO analysis of agency-provided data.

[a]At the time of our review, agency H had not submitted its September 2009 plan to DHS. This reported number is from an earlier plan that the agency provided to us.

Examples of the capabilities that agencies most frequently reported not having implemented included having secure facilities in place to handle classified information, being able to filter specific types of Internet traffic, and participating in the Einstein program.

Between October 2008 and September 2009, agencies delayed their plans for implementing the critical security capabilities. Of the 13 access provider agencies that had not implemented all of the required capabilities as of September 2009, 6 agencies delayed their expected planned dates for implementing the remaining critical capabilities between approximately 10 months and 3 years. As of September 2009, nine of these agencies were reporting that they expected to complete implementation of the remaining critical security capabilities between September 2009 and December 2010, one expected to complete its efforts in December 2013, and three did not project a date by which they expected to complete implementation.

Agencies Have Not Demonstrated Full Compliance with TIC Capabilities or Completed Consolidation Efforts

Agencies have not demonstrated full compliance with TIC capabilities. As of September 2009, DHS had conducted TIC Compliance Validation reviews at 6 of the 16 agencies in our review that are required to undergo a review, and the remaining 10 had been scheduled to be evaluated between October 2009 and May 2010.[24]

The results of the reviews indicated that information that agencies had reported was not always accurate. Specifically, although agencies had reported that certain capabilities were in place, the results for five of the six agencies that completed reviews indicated that several of these capabilities had not been fully implemented. For example, one agency's results showed that it had not fully implemented 10 critical capabilities, including 7 that it had previously reported as complete. In addition, the results for another agency showed that it had a large number of connections that it had not previously reported; the agency originally reported 119 connections, but after the review it identified 403 external connections. As indicated earlier, agencies are required to develop plans to address any shortcomings identified in the review and to submit their plans to DHS.

Agencies Experienced Benefits and Lessons Learned in Implementing TIC, but Challenges Remain in Complying with Requirements

While the TIC initiative offers benefits to agencies, such as improved network security, agencies have been challenged in complying with the requirements of the initiative, in part because of shortcomings in communication by OMB and DHS. In addition, agencies' experiences in implementing TIC offers valuable lessons learned for OMB and DHS that may increase the likelihood of the initiative's success.

Benefits in Improved Security and Network Management Are Anticipated

Although agencies are still in the process of implementing TIC, the initiative offers benefits to agencies.

Improved Network Security. TIC will improve security at agencies by reducing the number of access points that have to be monitored. Several agencies indicated that consolidating connections and centralizing security monitoring at TIC access points should make it easier to monitor traffic and protect their networks from attacks. In addition, officials from another agency stated that the consolidation of external connections had made the agency's network perimeter more secure.

Improved Network Management. The initiative has also helped improve agencies' management of their networks. Several agencies stated that implementing TIC by consolidating their external connections is beneficial because it has forced them to gain a greater awareness of their overall network environment. Another agency anticipated that TIC implementation would reduce complexity in its network, making it simpler to manage.

Agencies Faced Challenges with Implementing TIC Requirements

Agencies continue to face challenges in implementing TIC, including implementing the initiative with incomplete information about the number of access points for which they have been approved and about the technical security capabilities. Further, DHS will continue to face challenges in knowing whether the access points are adequately secured.

Implementing the initiative with incomplete information. Best practices for program management, established by the Project Management Institute in *The Standard for Program Management*,[25] state that the information that program stakeholders need should be made available in a timely manner throughout the life cycle of a program. In addition, our *Internal Control Management and Evaluation Tool* [26] states that when communicating with other agencies, managers should provide timely information that is relevant to the requester's needs. However, in some circumstances, agencies have been unable to effectively plan for implementing the initiative because OMB did not always consistently communicate the number of TIC access points for which agencies had been approved in a timely manner and DHS did not always promptly respond to agencies' questions about the required security capabilities.

OMB did not consistently inform agencies about the number of TIC access points for which they had been approved until more than a year after it required agencies to submit their requested number. In a memo issued in September 2009,[27] OMB announced that access provider agencies were each allowed two access points, 17 months after its April 2008 deadline for agencies to submit their requested number of trusted connections. However, between April 2008 and September 2009, OMB's communication of the number of access points it had approved for agencies was inconsistent. Specifically,

- Several agencies told us that OMB, or DHS rather than OMB, verbally told them about the number of access points for which they had been approved but did not provide them with written confirmation of the approved number.
- One agency said that it received an e-mail from DHS, as opposed to OMB, stating that its top two to three locations had been approved; however, officials from the agency indicated that the agency was not informed of the exact number of approved access points.
- A few other agencies stated that OMB never informed them of the number of approved access points, either verbally or in writing.

OMB addressed these shortcomings by issuing the memo in September 2009; however, any further inconsistencies in communication by OMB could cause additional challenges for agencies. In the memo, OMB also informed access provider agencies about the process for submitting an evidence-based rationale to DHS to request additional TIC access points. In this process, OMB is responsible for notifying agencies of its final decision on how many additional access points the agency is to be allowed. As described earlier, several agencies indicated that they planned to request additional access points. However, even with this process in place, agencies may still be uncertain about the number of access points for which they have been approved if prior inconsistencies in communication from OMB resurface. For example, although one agency's request for additional access points was sent to OMB in April 2009, as of December 2009 agency officials indicated that they not been told whether the agency's request had been approved. Without consistent and timely communication of the results of agency requests for additional access points by OMB, agencies that requested additional access points will continue to face challenges with implementation of TIC.

In addition, DHS often did not promptly respond to agency questions about the technical aspects of securing TIC access points, further complicating agency implementation efforts. Although a few agencies that have asked DHS questions about the meaning of specific terms in the security capabilities or about guidance for implementation stated that DHS answered their questions effectively, four agencies stated that DHS has often been slow to respond to questions about the capabilities, or in some cases has not responded at all. Specifically, one agency noted that DHS took a year to produce answers to frequently asked questions that were generated in an inter-agency working group. Three other agencies told us that they still have not received answers to questions that they submitted to DHS on specific security capabilities such as data storage requirements, inspection of encrypted traffic, and participation in the Einstein program. DHS officials acknowledged that its communications with agencies had not been timely because it had limited staff at the beginning of the initiative.

Without consistent and timely communication from OMB and DHS, agencies may not be able to effectively execute plans for consolidating their external connections and securing their TIC access points.

Ensuring that critical capabilities have been implemented. DHS will be challenged to know whether access providers have adequately secured their access points because it does not directly test the capabilities in its compliance validation reviews. The National Institute of Standards and Technology states[28] that organizations should conduct assessments to determine the extent to which controls are implemented correctly, operating as intended, and producing the desired outcome with respect to meeting the security requirements for the system. During its reviews, DHS conducts document reviews, interviews, and observation of agency processes, but does not conduct direct testing of the capabilities to determine if they are effectively implemented, operating as intended, and achieving desired results. Even with this limited testing, five of the six reviews that DHS conducted showed that agencies had not fully implemented critical security capabilities that had previously been reported as implemented. However, without directly testing the capabilities, DHS could be unaware of additional weaknesses that its more limited reviews may not have identified.

In addition, in at least three of the six reviews that it conducted at agencies, DHS did not evaluate all of the trusted connection locations. Specifically, in one agency's review, DHS evaluated only one of the agency's two security operations centers and one of its four TIC locations. According to DHS, the other center and three locations were not evaluated because the agency asserted that its other sites were identical to the ones evaluated. For another agency, DHS evaluated a security operations center and a telecommunications facility at the agency but did not examine controls at either of the agency's TIC access point locations. A third agency was only evaluated at one of its two TIC locations. DHS officials indicated that in designing the method for TIC compliance reviews, it was decided that the initial round of reviews would include only the most mature TIC locations and supporting network and operations centers. Without evaluating all agency locations in its compliance reviews, DHS cannot be assured that agencies have implemented critical capabilities at all locations.

Defining Requirements and Effective Communication Offer Lessons Learned for OMB and DHS as the Initiative Moves Forward

Agencies' experiences in implementing TIC offer valuable lessons learned for OMB and DHS.

Defining requirements clearly and early prove useful for agency planning. OMB and DHS did not always use sound program management principles when planning the TIC initiative. According to *The Standard for Program Management*, during the planning phase, program requirements should be developed before schedules are defined. However, OMB and DHS did not define certain fundamental requirements before establishing initial deadlines for the initiative. For example, DHS did not define the meaning of "external connection" until April 2009, 17 months after the initiative was announced and 10 months after the initial June 2008 deadline for reducing external connections to authorized levels. This resulted in DHS determining during a compliance validation review that one agency had not reported a number of external connections that needed to be consolidated. DHS officials acknowledged that this was due to confusion over the definition of what constituted an external connection. In addition, the technical security capabilities that would be required for access providers were still being defined when agencies developed their required initial implementation plans and were not finalized until April 2008, 5 months after the initiative was announced. As a result, several agencies stated that it was difficult for them to plan for TIC implementation. In going forward, defining any key future requirements prior to establishing deadlines will be critical to the initiative's success.

Collaborative meetings aided implementation. DHS and OMB sponsored several collaborative meetings during the initiative that many agencies found beneficial for their implementation of TIC. Specifically, several agencies stated that the meetings of the inter-agency TIC technical working group were helpful. For example, one agency said that DHS provided updates about the initiative during the meetings. Another agency noted that the meetings provided additional specificity on aspects of the program. Several agencies also stated that the meetings provided a forum for agencies to discuss issues related to TIC with one another, allowing them to gain insight from other agencies. One of these agencies found the meetings to be helpful because it was able to provide feedback to DHS about the technical capabilities. Another agency noted that it had recently participated in conference calls with DHS that helped to address its technical questions related to implementing the critical capabilities. In the future, continuing such effective communication increases the chances of the initiative's success.

Meeting business needs with a reduced number of connections is complex and time-consuming. As indicated earlier, the 16 access provider agencies in our review are reporting that they are reducing and consolidating from 3,286 external connections. Reducing to the approved total of 32 TIC access points is a complex and time-consuming effort for most agencies. For example, one agency indicated that implementing the infrastructure required to support its mission would require 4 years to complete. Two other agencies noted that implementing the initiative required them to make significant changes to their existing network architecture. In addition, for several agencies, determining how to meet their business needs within the technical constraints of TIC has been a complex task. For example, three agencies stated that they needed more than two TIC access points to ensure that their networks would remain operational in the event of a disaster. One of these agencies explained that its high performance and capacity requirements would not be met with only two access points. The complex effort required for agencies to implement the initiative while still meeting their business needs has led to significant delays in agencies' plans for implementation. As indicated earlier, the access provider agencies have reported that they have consolidated fewer connections than they originally planned and have significantly revised their future plans for consolidation. Recognizing that agencies may desire more than two access points, as noted earlier, OMB and DHS established a process for agencies to submit an evidence-based rationale for obtaining additional access points.

DHS Has Deployed Einstein to Six Agencies, but Faces Challenges with Meeting Program Goals

Einstein is intended to provide DHS with an increased awareness of activity, including possible security incidents, on federal networks. As of September 2009, fewer than half of the 23 agencies had executed the required agreements with DHS, and Einstein 2 had been deployed to six agencies. Agencies that participated in Einstein 1 improved identification of incidents and mitigation of attacks, but DHS continues to face challenges with meeting the goals of the initiative.

Einstein Is to Provide Increased Awareness of Activity on Agency Networks

The goal for Einstein is to provide US-CERT with a higher level of awareness of activity on federal networks. By implementing this initiative, DHS intended to achieve the following objectives:

- provide an automated process for collecting, correlating, and analyzing computer network traffic information from participating federal agencies;

- provide US-CERT with a means to observe potential malicious activity in computer network traffic entering and exiting participating agencies' computer networks;
- increase US-CERT's situational awareness of federal agency computer networks through correlation of activity across the entire federal enterprise; and
- incorporate intrusion detection technology (i.e., the Einstein sensors and signature-monitoring capabilities) capable of alerting US-CERT to the presence of malicious or potentially harmful computer network activity in federal agencies' network traffic.

DHS and Agencies Are Required to Take Various Actions before Einstein 2 Can Be Deployed

To accomplish these objectives, for Einstein 2, agencies are required to meet the following two requirements:[29]

- **Execute a memorandum of agreement with DHS.** This agreement establishes the responsibilities of deployment and operation of the sensor between the participating federal agency and DHS.
- **Execute a service level agreement with DHS.** This agreement defines the roles, responsibilities, and points of contact, as well as describes the services, hours of operation, and performance levels provided to the agency. It also requires agencies to update US-CERT regularly on the status of ongoing investigations related to alerts.

  Agencies were required to report on the status of these agreements to DHS in September 2009.

  In addition, the TIC access provider agencies are required to meet two additional requirements:
- **Execute an interconnection security agreement with DHS.** Describes the interconnection between the agency and DHS and the security controls required and implemented to protect the confidentiality, integrity, and availability of the systems and data. Agencies were required to report on their status in completing this agreement to DHS in September 2009.
- **Perform a site assessment.** Provides a technical description of the agency's network and how the network connects to the agency's Internet service providers.

Vendors that intend to provide TIC services to agencies under the Networx contract are also required to complete a memorandum of agreement, an interconnection security agreement, and a site assessment.

With the required agreements in place,[30] DHS is to deploy Einstein sensors to access provider agencies and Networx vendors. When deploying the sensors, DHS is to use a site deployment checklist to verify that the Einstein equipment is installed and configured appropriately. After the sensors are operational, US-CERT is to begin monitoring and analyzing results.

Einstein 2 Has Been Deployed to Six Agencies, but DHS and Agencies Did Not Always Complete Required Activities

As of September 2009, DHS had deployed Einstein 2 at six access provider agencies included in our review and at three Networx vendors. According to DHS, the sensors at five of the six agencies were operational as of September 2009; it had not activated the sensors at one agency because it was waiting for the agency to complete required agreements.

Agencies that had operational sensors had completed certain required agreements, but not all agencies had executed all required agreements. All five agencies with operational sensors had executed memorandums of agreement and interconnection security agreements with DHS as required. However, three of the five agencies had not executed service level agreements. According to DHS officials, these agencies were still in the process of negotiating the agreements. However, the agreements define key requirements for the initiative, including how US-CERT is to notify agencies of potential incidents and how agencies are to respond to these notifications, including what information must be provided to US-CERT in support of investigations related to Einstein alerts. Without these agreements in place between agencies and DHS, agencies may not receive the information needed to address security incidents detected by Einstein, and DHS may not obtain the information it needs from agencies in order to fully meet the objective of improving situational awareness.

DHS and the agencies also did not always complete deployment checklists. Although all five of the agencies had performed required site assessments, the site deployment checklists for two agencies had not been signed by officials from the agency or from DHS verifying that the sensors had been installed and configured appropriately. As a result, DHS and agency management cannot be assured that the Einstein equipment has been installed and configured appropriately.

Because these sensors had only recently been deployed, we did not evaluate the extent to which US-CERT was collecting and analyzing data and reporting alerts to agencies for Einstein 2.

Not all of the remaining 17 agencies reported their status toward submitting required agreements to DHS in September 2009. Only a few have reported completing required agreements with DHS, while several have not yet reported their plans for submitting agreements.[31] Specifically:

- Four agencies reported that they had completed and submitted their memorandums of agreement to DHS, and 4 reported that they expected to submit them within a year; however, nine did not project a date by which they expected to submit them.
- One agency reported that it had submitted its service level agreement to DHS, and 4 reported that they expected to submit them between December 2009 and September 2010; however, 12 did not project a date by which they expected to submit the agreement.
- Two of the 10 remaining agencies required to execute interconnection security agreements[32] reported that they had submitted them to DHS, and 1 reported that it expected to submit the agreement within the next year; however, 7 did not project a date by which they expected to submit the agreements.

Although DHS required agencies to report their status toward executing required agreements in September 2009, it did not establish milestones for agencies to submit the agreements. According to *The Standard for Program Management*, the actual completion of program activities and milestones should be tracked against a planned timeline in order to ensure that the program produces its required deliverables on time. However, DHS had not established any milestones for agencies to submit these agreements. As indicated earlier, these agreements establish key responsibilities and controls that are necessary for successful operation of the sensors. Without establishing milestones for these agreements, DHS could face delays in deploying and activating Einstein sensors.

Einstein Has Proven Beneficial to Providing Security, but DHS Faces Ongoing Challenges with Meeting Program Goals

Agencies have benefited from Einstein alerts, and their experiences have provided DHS with valuable lessons; however, DHS may be challenged in meeting program goals as the system is deployed at more agencies.

Einstein Provided Security Benefits for Agencies

Although Einstein 2 has only been deployed at 6 agencies, the 12 agencies that participated in Einstein 1 realized benefits in the following areas:

Identifying incidents. US-CERT provided alerts to agencies from its analysis of the data from the Einstein 1 sensors, which contained information about potential cyber attacks or incidents against the agency's networks. Several agencies observed that the alerts from US-CERT were helpful or contained useful information about potential incidents, including information that could be used to trace potential incidents to specific locations on the network. For some agencies, Einstein identified incidents that agencies' intrusion detection systems had not found, increasing their ability to mitigate potential attacks.

Providing cross-agency view. For Einstein 1, US-CERT provided reports based on a correlation of sensor data from all of the participating agencies. Several agencies said US-CERT's ability to aggregate Einstein data from multiple agencies was beneficial for identifying potential attacks against government networks.

Using sensor data. In addition to receiving alerts generated by US-CERT's analysis, agencies had the ability to access the sensor data directly via a Web portal. Several agencies indicated that they used this data to look for potential incidents on their own.

DHS Faces Challenges with Meeting Einstein Goals and Providing Adequate Analysis

As DHS deploys Einstein across the government, it faces the following challenges:

Understanding whether alerts are valid. Although one of the objectives of Einstein is to improve situational awareness of activity across the federal government, DHS will be challenged in understanding the extent to which this objective is being met because it lacks performance measures for Einstein 2 that address whether or not agencies report that the alerts represent actual incidents. For Einstein 1, agencies did not always inform US-CERT of how they responded to the alerts. As a result, US-CERT did not know whether these alerts represented false positives or actual incidents. We have previously reported that performance measures are most meaningful when they are linked with organizational goals.[33] DHS's performance measures for Einstein 2 indicate the time required for the system to detect known cyber events and to generate automated notifications once the events are detected, but they do not indicate agencies' responses to alerts. Establishing such measures would help DHS better understand whether the alerts are valid, helping it to better determine the extent to which the initiative is meeting its objective of improving situational awareness.

Having staff with required skills to monitor and analyze data. DHS will be challenged to have staff with the appropriate skills to fulfill its analysis and incident response mission as Einstein 2 is deployed across the government. As more agencies receive sensors, the amount of data that US-CERT will be responsible for analyzing will drastically increase. DHS recognizes that staff with appropriate analytical skills will be required in order to handle the increased workload, but it has not developed a staffing plan to address its need to acquire and retain qualified analysts at US-CERT. Although the department announced in October 2009 that it plans to hire up to 1,000 new cybersecurity professionals over the next 3 years, we previously reported in July 2008 that obtaining and retaining adequately trained cyber analysts is an ongoing challenge to US-CERT that hinders its ability to respond to increasingly fast, nimble, and sophisticated cyber attacks. At that time, we recommended that the department address the challenges that have impeded it from expeditiously hiring sufficiently trained cyber analysts and developing strategies for hiring and retaining highly qualified cyber analysts.[34] Although DHS indicated that it plans to expedite the hiring and on- boarding process for new analysts and to offer appropriate training opportunities for its analysts, it has not yet provided evidence that it has taken these actions. Until DHS addresses our prior recommendation by developing strategies for hiring and retaining cyber analysts, US-CERT may lack staff with appropriate skills to analyze the Einstein data, increasing the risk that attacks against federal networks could go undetected.

Additional Information from US-CERT Helped Agencies, Providing Valuable Lessons Learned

Agencies' experiences with the initial version of Einstein provided DHS with lessons learned for future versions of the initiative.

Detailed and timely information from alerts proved useful. Several agencies' experiences with Einstein 1 improved over time because information provided by US-CERT increased in its timeliness and detail. Although some agencies said that the alerts and reports that US-CERT provided were not always timely and useful, a few agencies observed that the information had improved over time. For example, one agency stated that the alerts lacked sufficient contextual information, making it difficult to determine whether the alerts were identifying false positives or actual incidents; however, several agencies indicated that the alerts had since improved in their usefulness. In addition, although several agencies noted that the alerts were not very timely when the sensors were first installed, a few indicated that the timeliness had improved for more recent alerts. Going forward, continuing to provide appropriate and timely information from the alerts will prove useful for agencies.

Access to sensor data proved useful for agencies. Further, several agencies that had direct access to the flow records from the Einstein sensor found that it was helpful in detecting potential incidents. DHS stated that all agencies participating in Einstein 2 will also have access to the flow data, which could provide similar benefits. However, not all agencies were aware that they would have access to this data. Making them aware of this and of the data's possible benefits could aid agencies in improving their monitoring of potential incidents.

CONCLUSIONS

TIC and Einstein are ambitious efforts that can help improve security and situational awareness across the federal government. However, in implementing the initiatives, federal agencies have faced challenges. For TIC,

OMB did not consistently communicate the number of access points for which agencies had been approved, and DHS did not always provide timely answers to agency questions about technical capabilities. In addition, because DHS does not conduct direct testing of the capabilities or evaluate all possible locations in its validation reviews, it cannot be assured that all critical capabilities have been implemented. For Einstein, the initiative could fail to fully meet the objective of increasing US-CERT's situational awareness because DHS did not always ensure that key agreements were executed with agencies. DHS could also be challenged in determining whether the initiative is meeting this objective without performance measures that indicate whether the alerts provided to agencies represent actual incidents. Without improvements in program management and communication from OMB and DHS, federal agencies will continue to be faced with challenges in implementing these initiatives that could ultimately jeopardize their ability to reduce and secure Internet connections.

With agencies still in the process of implementing TIC and DHS in the early stages of deploying Einstein 2, the success of such large-scale initiatives will be in large part determined by the extent to which DHS, OMB, and other federal agencies work together to address the challenges of these efforts and to apply lessons learned during the initial stages of implementation. Although this will not guarantee the success of TIC and Einstein, doing so will enhance the chances that the initiatives will meet their goals of reducing, consolidating, and securing federal Internet connections.

RECOMMENDATIONS FOR EXECUTIVE ACTION

In order to ensure that federal agencies continue to have adequate information about the number of connections for which they have been approved, we recommend that the Director of OMB take the following two actions:

- Communicate its final decisions on agency requests for additional TIC access points in a consistent and timely manner.
- Assess the efficacy of, and take steps to apply as appropriate, the lesson learned during the initial implementation of TIC regarding the need to define future requirements before establishing deadlines.

In addition, in order to further ensure that federal agencies have adequate, sufficient, and timely information to successfully meet the goals and objectives of the TIC and Einstein programs, we recommend that the Secretary of Homeland Security take the following six actions:

- Provide agencies with timely responses to their questions seeking clarification on TIC security capabilities.
- Enhance TIC compliance validations by including (1) direct testing and evaluation of the critical capabilities and (2) evaluation of the capabilities at all agency TIC locations.
- Before activating Einstein sensors, ensure that both DHS and participating agencies (1) execute required service level agreements and (2) sign site deployment checklists.
- Establish milestones for agencies to submit required Einstein agreements.
- To better understand whether Einstein alerts are valid, develop additional performance measures that indicate how agencies respond to alerts.
- Assess the efficacy of, and take steps to apply as appropriate, lessons learned during the initial implementation of these initiatives such as the need to (1) define future requirements for TIC before establishing deadlines and (2) make agencies aware of their ability to access Einstein flow data.

AGENCY COMMENTS AND OUR EVALUATION

We provided a draft of this report to OMB and DHS for their review and comment. In providing e-mail comments on a draft of this report, the lead information technology policy analyst from OMB's Office of E-Government and Information Technology stated that OMB concurred with the report's findings, conclusions, and two recommendations addressed to OMB. In email comments provided by an audit liaison from DHS's Office of Cybersecurity and Communications, DHS concurred with the six recommendations addressed to DHS. DHS also provided technical comments, which we have incorporated into this report as appropriate. We also provided a draft of this report to the 22 other agencies included in our review. Of the 22, 15 responded that they did not have any comments; 1 provided technical comments, which we addressed as appropriate; and 6 did not respond.

Gregory C. Wilshusen
Director, Information Security Issues

APPENDIX I. OBJECTIVES, SCOPE, AND METHODOLOGY

The scope of our review covered two initiatives: Trusted Internet Connections (TIC) and the National Cybersecurity Protection System (NCPS) program, operationally known as Einstein. For each initiative, our objectives were to (1) identify their goals, objectives, and requirements; (2) determine the status of the actions federal agencies have taken, or plan to take, to implement them; and (3) identify the benefits, challenges, and lessons learned in implementing each initiative.

For TIC, to address the first objective, we obtained and reviewed applicable policies and memorandums issued by the Office of Management and Budget (OMB) and guidance, reports, and other documentation provided by the Department of Homeland Security (DHS). We also held discussions with OMB and DHS representatives concerning the goals, objectives, and requirements of the initiative. To understand the options for agencies seeking to acquire TIC services through the Networx contract, we obtained and reviewed relevant documents regarding Networx and interviewed officials from the General Services Administration.

To address the second objective for TIC, we reviewed statements of capability, plans of action and milestones, and other relevant documents for 23 of the 24 agencies[35] listed in the Chief Financial Officers Act of 1990[36] to determine if reporting requirements were met. We also reviewed these documents to determine reported progress toward the reduction and consolidation of external connections and implementation of critical capabilities and analyzed them to estimate the overall progress reported by agencies. We also reviewed documentation from DHS to determine whether agencies submitted the required documents. In addition, we reviewed the results of six TIC Compliance Validation reviews and interviewed officials from DHS to understand how the department assesses agencies' degree of compliance with TIC and to determine the extent to which the information reported in agency plans of action and milestones was accurate.

To address the third objective for TIC, we interviewed officials from each agency, DHS, and OMB. In addition, we obtained written responses to follow-up questions from each agency. We also examined plans of action and milestones and other relevant documents from each agency and reviewed policies and guidance from OMB and DHS to identify any additional benefits, challenges, or lessons learned. Further, we interviewed officials from agency inspectors general to obtain information on any benefits, challenges, or lessons learned that they had identified related to the initiative.

For Einstein, to address the first objective, we obtained and reviewed applicable policies, guidance, and other documentation provided by DHS. We also held discussions with DHS officials concerning the goals, objectives, and requirements of the initiative.

To address the second objective for Einstein, we reviewed plans of action and milestones for each agency to determine whether reporting requirements were met. In addition, we examined required agreements and site assessments for the six agencies where Einstein 2 was deployed to verify their completion. We also interviewed officials and obtained written information from DHS and from each agency to obtain additional information on the status of implementation.

To address the third objective for Einstein, we interviewed officials from DHS and from each agency. In addition, we obtained and reviewed written responses to follow-up questions from each agency. We also examined policies, guidance, and other documentation from DHS to identify any additional benefits, challenges, or lessons learned. Further, we interviewed officials from agency inspectors general to obtain information on any benefits, challenges, or lessons learned that they had identified related to the initiative.

We conducted this performance audit from December 2008 to March 2010 in Washington, D.C., in accordance with generally accepted government auditing standards. Those standards require that we plan and perform the audit to obtain sufficient, appropriate evidence to provide a reasonable basis for our findings and conclusions based on our audit objectives. We believe that the evidence obtained provides a reasonable basis for our findings and conclusions based on our audit objectives.

End Notes

[1] Director of National Intelligence, *Annual Threat Assessment of the Intelligence Community for the Senate Select Committee on Intelligence*, statement before the Senate Select Committee on Intelligence (Feb. 12, 2009).

[2] The Comprehensive National Cybersecurity Initiative consists of 12 projects intended to improve DHS's and other federal agencies' efforts to safeguard federal executive branch government information systems by reducing potential vulnerabilities, protecting against intrusion attempts, and anticipating future threats against the federal government's networks.

[3] The 24 agencies subject to the act are the Departments of Agriculture, Commerce, Defense, Education, Energy, Health and Human Services, Homeland Security, Housing and Urban Development, the Interior, Justice, Labor, State, Transportation, the Treasury, and Veterans Affairs; the Environmental Protection Agency, General Services Administration, National Aeronautics and Space Administration, National Science Foundation, Nuclear Regulatory Commission, Office of Personnel Management, Small Business Administration, Social Security Administration, and U.S. Agency for International Development.

[4] GAO, *High-Risk Series: An Update*, GAO-09-271 (Washington, D.C.: January 2009).

[5] GAO, *Information Security: Agencies Continue to Report Progress, but Need to Mitigate Persistent Weaknesses*, GAO-09-546 (Washington, D.C.: July 17, 2009).

[6] The Federal Information Security Management Act was enacted as title III, E-Government Act of 2002, Pub L. No. 107-347, 116 Stat. 2899, 2946 (Dec. 17, 2002).

[7] GAO, *Information Security: Progress Reported, but Weaknesses at Federal Agencies Persist,* GAO-08-571T (Washington, D.C.: Mar. 12, 2008).

[8] GAO, *Cybersecurity: Progress Made but Challenges Remain in Defining and Coordinating the Comprehensive National Initiative*, GAO-10-338 (Washington, D.C.: Feb. 1, 2010).

[9] OMB, *Implementation of Trusted Internet Connections (TIC)*, M-08-05 (Washington, D.C.: Nov. 20, 2007).

[10] According to DHS officials, each authorized TIC access point may include one or more external connections.

[11] Examples of connections that are not required to be routed through an approved TIC include (1) dedicated connections to agency remote offices that do not pass through the Internet, (2) connections made using technology that provides a secure communication mechanism for data transmitted across public networks (i.e., virtual private networks), and (3) connections with other agencies where both agencies have implemented TIC.

[12] Although OMB originally designated 17 of the 23 agencies in our review as TIC access providers, one of these agencies has since chosen to seek service from another access provider.

[13] According to DHS officials, in December of 2008, the Einstein program was incorporated into NCPS, a larger collection of systems that includes not only the Einstein sensors, but also other systems providing data correlation and analysis.

[14] Established by DHS, the US-CERT serves as a focal point for the government's interaction with federal and nonfederal entities on a 24-hour-a-day, 7-day-a-week basis regarding cyberrelated analysis, warning, information sharing, major incident response, and national-level recovery efforts. It is charged with aggregating and disseminating cybersecurity information to improve warning of and response to incidents, increasing coordination of response information, reducing vulnerabilities, and enhancing prevention and protection. In addition, US-CERT collects incident reports from all federal agencies and assists agencies in their incident response efforts.

[15] Network flow records are records of communications made to an organization's IT systems. The records identify the source and destination Internet Protocol addresses used in the communication, the source and destination ports, the time the communication occurred, and the protocol used to communicate.

[16] Signatures are recognizable, distinguishing patterns associated with cyber attacks, such as a binary string associated with a computer virus or a particular set of keystrokes used to gain unauthorized access to a system.

[17] When the initiative was first announced in November 2007, OMB set a target number of 50 connections across the federal government. However, OMB officials have since stated that the target number is no longer applicable and that a new target has not been established.

[18] OMB, *Guidance for Trusted Internet Connection Statement of Capability Form (SOC)*, M-08-16 (Washington, D.C.: Apr. 4, 2008).

[19] The seven agencies in our review that are seeking service from other providers were not authorized a specific number of access points.

[20] As of September 2009, six access provider agencies were targeting more access points than the number for which they had been approved by OMB.

[21] At the time of our review, one access provider agency had not submitted its September 2009 progress report to DHS; the status of its consolidation effort, reflected in the table, is based on its July 2009 progress report.

[22] Seeking service agencies are not included in this figure.

[23] The one multi-service access provider agency reported that it had implemented all of the 11 additional critical security capabilities required for multi-service access providers.

[24] According to DHS officials, only one of the four participating Networx vendors had passed a review and could offer TIC services to agencies.

[25] Project Management Institute, *The Standard for Program Management, Second Edition* (Newton Square, Pa.: 2008).

[26] GAO, *Internal Control Management and Evaluation Tool*, GAO-01-1008G (Washington, D.C.: August 2001).

[27] OMB, *Update on the Trusted Internet Connections Initiative*, M-09-32 (Washington, D.C.: Sept. 17, 2009).

[28] National Institute of Standards and Technology: *Recommended Security Controls for Federal Information Systems*, Special Publication 800-53 Revision 3 (Gaithersburg, Md.: December 2007).

[29] For Einstein 1, DHS required participating agencies to complete a memorandum of agreement, interconnection security agreement, and a site assessment before receiving a sensor.

[30] Although agencies are required to complete a service level agreement, DHS officials stated that it is not necessary for it to be completed before the Einstein sensors are deployed.

[31] One access provider agency did not submit its updated plan to DHS in September 2009.

[32] The seven agencies seeking service from other access providers are not required to execute interconnection security agreements.

[33] GAO, *Information Security: Concerted Effort Needed to Improve Federal Performance Measures*, GAO-09-617 (Washington, D.C.: Sept. 14, 2009).

[34] GAO, *Cyber Analysis and Warning: DHS Faces Challenges in Establishing a Comprehensive National Capability*, GAO-08-588 (Washington, D.C.: July 31, 2008).

[35] The Department of Defense was not included in our review because it was not required to implement TIC or Einstein.

[36] 31 U.S.C. §901(b).

CHAPTER SOURCES

Chapter 1 - This is an edited, reformatted and augmented version of a U.S. Government Accountability Office publication, GAO-11-43, dated November, 2010.

Chapter 2 - This is an edited, reformatted and augmented version of a U.S. Government Accountability Office publication, GAO-11-29, dated November, 2010.

Chapter 3 - This is an edited, reformatted and augmented version of a U.S. Government Accountability Office publication, GAO-11-67, dated December, 2010.

Chapter 4 - This is an edited, reformatted and augmented version of a U.S. Government Accountability Office publication, GAO-10-237, dated March, 2010.

INDEX

A

B

C

D

N

O

P

Q

R

S

T